# Advances in Chromatography

# Advances in Chromatography

## Volume 56

Edited by

Nelu Grinberg

Peter W. Carr

CRC Press is an imprint of the
Taylor & Francis Group, an **informa** business

ISSN 0065-2415

CRC Press
Taylor & Francis Group
6000 Broken Sound Parkway NW, Suite 300
Boca Raton, FL 33487-2742

First issued in paperback 2022



ISBN 13: 978-1-03-240164-5 (pbk)
ISBN 13: 978-0-367-13375-7 (hbk)

DOI: 10.1201/9780429026171



Publisher's Note
The publisher has gone to great lengths to ensure the quality of this reprint but points out that some imperfections in the original copies may be apparent.

**Visit the Taylor & Francis Web site at**
**http://www.taylorandfrancis.com**

**and the CRC Press Web site at**
**http://www.crcpress.com**

# Contents

# Editors

**Nelu Grinberg** recently retired following a career in research chemistry in the pharmaceutical industry. His research in analytical chemistry has an emphasis on chromatography, spectroscopy, and chiral separations. Until January 2017, Dr. Grinberg was a Distinguished Research Fellow in the Chemical Development Department at Boehringer Ingelheim Pharmaceuticals in Ridgefield, Connecticut, prior to which, he worked for sixteen years as a Senior Research Fellow in the Analytical Department at Merck Research Laboratories in Rahway, New Jersey. Dr. Grinberg has authored and co-authored over 150 publications, including articles and book chapters, and has lectured and conducted courses worldwide. He is currently editor-in-chief of the *Journal of Liquid Chromatography and Related Techniques*, editor of the Chromatographic Science book series, editor of the Supramolecular book series, and co-editor of the Advances in Chromatography series. He is also a member of the board of the Connecticut Separation Science Council and was a recipient of a Koltoff Fellowship of the Hebrew University of Jerusalem. Dr. Grinberg obtained his PhD in chemistry from the Technical University of Lasi in Romania. He conducted post-doctoral research with Professor Barry Karger at Northeastern University in Boston, Massachusetts, and with Professor Emanuel Gil-Av at the Weizmann Institute of Science in Rehovot, Israel.

**Peter W. Carr** is currently a professor of chemistry at the University of Minnesota (Twin Cities), which he joined in 1977 after being on the faculty of the University of Georgia (Athens) since 1969. He was educated at the Polytechnic Institute of Brooklyn (BS 1965) and the Pennsylvania State University (PhD 1969), simultaneously holding post-doctoral appointment at Stanford University Medical School (1968–1969). His major interests have touched on electroanalytical chemistry, thermal analysis and chemical thermodynamics, and most recently separation science. He has received numerous awards in analytical chemistry (Benedetti Pichler Award, Pittsburgh Award in Analytical Chemistry, and American Chemical Society Award in Analytical Chemistry) as well as various awards in chromatography (EAS, ACS, Dal Nogare Award of the Chromatography Forum of the Delaware Valley, Martin Medal of the Chromatographic Society, and Horvath Medal of the Connecticut Chromatography Society). He has published over 400 papers and book chapters. His major research contributions in separation science include retention mechanisms in gas, liquid, and supercritical fluid chromatography (especially reversed phase chromatography), development of ultra-stable phases for liquid chromatography (LC), high speed LC, high temperature LC, theory of peak capacity in gradient LC, and high speed 2D LC. He was a founding father of the Minnesota Chromatography Forum and its first president.

# Contributors

**Alain Beck**
IRPF, Center of Immunology Pierre Fabre
Saint-Julien-en-Genevois, France

**Szymon Bocian**
Department of Environmental Chemistry & Bioanalytics
Faculty of Chemistry
Nicolaus Copernicus University
Toruń, Poland

**Yannis-Nicolas François**
Laboratoire de Spectrométrie de Masse des Interactions et des Systèmes (LSMIS) UMR 7140 (Unistra-CNRS)
Université de Strasbourg
Strasbourg, France

**Rabah Gahoual**
Faculté de Pharmacie de Paris
Unité de Technologies Chimiques et Biologiques pour la Santé (UTCBS)
University Paris Descartes
Paris, France

**Anne-Lise Marie**
Department of Chemistry and Chemical Biology
Barnett Institute of Chemical and Biological Analysis
Northeastern University
Boston, Massachusetts

**Paul Nelson**
Ramsey County Historical Society
University of Minnesota
St. Paul, Minnesota

**Coralie Ruel**
Institut Galien Paris Sud
Université Paris-Sud, CNRS
Université Paris-Saclay
Châtenay-Malabry cedex, France

**Joseph Sherma**
Department of Chemistry
Lafayette College
Easton, Pennsylvania

**Gregory O. Staples**
Agilent Technologies
Santa Clara, California

**Dwight R. Stoll**
Department of Chemistry
Gustavus Adolphus College
St. Peter, Minnesota

**Myriam Taverna**
Institut Galien Paris Sud
Université Paris-Sud, CNRS
Université Paris-Saclay
Châtenay-Malabry cedex, France

**Nguyet Thuy Tran**
Institut Galien Paris Sud
Université Paris-Sud, CNRS
Université Paris-Saclay
Châtenay-Malabry cedex, France

**Kelly Zhang**
Genentech
South San Francisco, California

# 1 Kolthoff of Minnesota

*Paul Nelson*

## CONTENTS



## FOREWORD

Izaak Maurits Kolthoff was doubtlessly the preeminent analytical chemist of the twentieth century—certainly in the United States and perhaps in the world. Educated in the Netherlands, Kolthoff adopted as his own the scientific motto of his esteemed mentor: "*Theory guides; experiment decides*." Kolthoff's mentor Professor Nicholas Schoorl had obtained his degree studying with the great Jocobus van't Hoft who was the first Nobel Laureate in Chemistry (1901). The influence of his scientific background in physical chemistry is clear both in Kolthoff's research and his teaching. His life's work became the transformation of a highly descriptive empirical branch of chemistry, namely chemical analysis, into the science of analytical chemistry, building on the physicochemical principles and concepts introduced by the founders of physical chemistry (such as Arrhenius and Ostwald). Although not a practitioner of chromatography, Kolthoff was the chief architect of much of the foundational chemical science upon which analytical chemistry and separation science are built.

Professor Kolthoff published his first paper in 1915. It concerned acid–base titrations using the recently developed theories of weak and strong electrolytes of Arrhenius and Sorenson's concept of pH. Because of his worldwide reputation ensued over the next 18 years, he joined the University of Minnesota in 1928, where he remained a faculty member until his so-called "retirement" in 1962. He published more than 800 papers during the "active" years of his career, and an additional 150 papers appeared before his health failed. He maintained a National Science Foundation grant until only two years before his passing.

Professor Kolthoff's research, covering a dozen areas of chemistry, focused primarily on constructing a firm scientific foundation for analytical chemistry. He and his students studied basic aspects of acidimetry, alkalimetry, establishing the pH scale using acid–base indicators, gravimetric analysis, iodometry, the theory of colloids and adsorption (especially important to chromatography) and crystal growth, thereby establishing the scientific basis for gravimetry. His work with acids and bases, in which he was the first to fully apply the Arrhenius theory (1884), developed the rational choice of end point indicators and transformed titrimetry from a practical art to a quantitative science based on sound principles. Kolthoff's work in acid–base chemistry—both in aqueous and non-aqueous media—is of seminal importance throughout the science of chemical separations. He developed the theory of potentiometric analysis and potentiometric titrations, as well as conductometric titrations. When Jaroslav Heyrovsky, at the Charles University (Prague, Czechoslovakia), discovered polarography (1925), for which he received the Nobel Prize (1959), Kolthoff immediately recognized both its scientific significance and practical importance. Indeed, it eventually led to important methods of environmental trace metal analysis and biological sensors in use today; it is also the basis of amperometric detectors, which are vital to chromatographic detection of neurotransmitters in analytical neurochemistry. Kolthoff's fundamental studies of diffusion and mass transfer as they pertain to electrochemistry and electroanalytical chemistry account, in part, for why so many of his students and their progeny became interested in chromatographic science. He and his students, especially James J. Lingane (PhD, 1938, Harvard University) and Herbert A. Laitinen (PhD, 1940, University of Illinois and Florida), contributed significantly to the understanding and use of electroanalytical methods such as polarography and ion-selective electrodes.

Throughout his career Kolthoff emphasized the role of chemico-physico principles in chemical analysis. Thus he was one of the earliest to understand the fundamental significance of the chemistry of crown ethers and cryptands and recognized the work of Jean Marie Lehn, who ultimately received the Nobel Prize in Chemistry (1987) for his work with cryptands. Lehn was the inaugural Kolthoff Lecturer at the University of Minnesota (1979).

In addition to his phenomenal production of research publications, Kolthoff wrote and edited many books, including the multivolume monograph Volumetric *Analysis* (with Vernon Stenger, PhD, 1933); *Polarography* with J. J. Lingane; and *Potentiometric Titrations* with H. A. Laitinen. Kolthoff was also editor-in-chief of the *Treatise on Analytical Chemistry* (19 volumes in 2 editions) and the inaugural editor of the series of monographs entitled "*Chemical Analysis*," which continues till this day. Many of the *Chemical Analysis* monographs are devoted solely to separation science.

Arguably his single most influential book was "*Quantitative Inorganic Analysis*," co-authored with his first doctoral student, Ernest B. Sandell (PhD, 1932). Widely recognized as the progenitor of all modern textbooks on analytical chemistry, it appeared in four editions (the last co-authored with colleagues Sandell, Meehan and Bruckenstein) and six languages (including Russian and Japanese).

His myriad contributions were recognized throughout the world. Being a member of the National Academy of Sciences, he received the American Chemical Society's Nichols Medal and the Fisher Award for Analytical Chemistry. He was the first recipient of the Kolthoff Award of the American Pharmaceutical Association.

His work with the most direct public impact came during the Second World War. Arriving in the United States from the Netherlands, where his family was virtually wiped out during the Nazi occupation, Kolthoff quickly assembled an extensive research group that made major contributions to the synthetic rubber program. He and his coworkers held several key patents related to synthetic rubber.

Chapter 1 in this volume is virtually unique in that it focuses more on Kolthoff's non-professional life, about which very little has been written compared to the extensive papers, interviews, and summaries emphasizing his profound scientific contributions. Perhaps the most detailed of these is the National Academy of Science memorial on Kolthoff written by his student Professor Johannes Coetzee (PhD, 1957), late of the University of Pittsburgh, Pittsburgh, Pennsylvania.

As Paul Nelson details below, following the war, Kolthoff was deeply engaged in many humanitarian efforts. He was an early supporter of the United Nations, and as a promoter of world peace, he corresponded extensively with Albert Einstein, Pierre Joliot-Curie, Senator Hubert Humphrey, and Mrs. Eleanor Roosevelt. He was also an early critic of Sen. Joseph McCarthy.

Many of his graduate students, including Sandell, Lingane, and Laitinen, went on to very successful careers in industry and academic life at institutions such as Harvard University, MIT, Northwestern University, the University of Chicago, the University of Michigan, the University of Pittsburgh, Pennsylvania State University, the University of Illinois, the University of California (Berkeley), and other major US research centers. By 1982 more than 1,100 PhDs could trace their scientific roots to him. The number of undergraduate students who learned from him, his students, or his books is incalculable. He stated to this author that the recognition that gave him the greatest pleasure was the American Chemical Society Division of Analytical Chemistry's inaugural Award for Excellence in Education, in 1983. In the words of James Lingane, "[A]nalytical chemistry has never been served by a more original mind, nor a more prolific pen, than Kolthoff's."

It was my honor and privilege to be a third-generation Kolthoff PhD, and subsequently to become his colleague while he was emeritus professor at Minnesota, where friends and colleagues knew him simply as "Piet," a shortening of his childhood nickname Pietje (roughly "little fellow"). We met for lunch and a chat almost weekly and sometimes on weekends. We discussed his many interests and reminiscences. Through him I met many of the international leaders in analytical chemistry, including chromatographers such as Phyllis Brown, Les Ettre, and Al Zlatkis, who, whenever in the Midwest, made a pilgrimage to Minnesota to visit with Piet Kolthoff.

**Peter W. Carr**
*Professor of Chemistry*
*University of Minnesota*
*July 9, 2018*

## 1.1 INTRODUCTION

On August 5, 1927, Dr. Samuel Lind, dean of the Institute of Technology at the University of Minnesota, sent this cablegram:

> University of Minnesota Minneapolis desires full professor analytical chemistry. Kruyt recommends you. Would you accept visiting professorship for coming year October to June for $4500 with permanent position in view? These few lines yielded riches for the university, in honors, teaching, scientific achievement, and, yes, money, beyond anyone's imagining.

The recipient was Izaak Maurits Kolthoff, a 33-year-old lecturer at the University of Utrecht, The Netherlands. Kolthoff did not know Lind, had never been to Minnesota, and might have been surprised to learn that a somewhere university bore that peculiar name. Three days later, Kolthoff cabled back his acceptance. Thus began a relationship that lasted 66 years.[1]

Kolthoff in laboratory, 1940s. University of Minnesota Archives, University of Minnesota—Twin Cities.

## 1.2 THE CHEMIST

"Don't worry, mama, I will restore your chicken soup to its optimal pH." Such is the gist of a story that Piet (his nickname from childhood, pronounced Pete) Kolthoff liked to tell and that he told often. He was 15 years old and had converted part of the family kitchen into a makeshift chemistry lab.

> One day I got in the kitchen and found my mother desolate. By mistake Mother had put in the chicken soup ... several large spoonfuls of sodium carbonate [baking soda] instead of sodium chloride [table salt.] She was just ready to throw everything into the sink when I told her that it was child's play to transform the carbonate into sodium chloride. Thus, I made my first titration, adding hydrochloric acid until—at a pH of 7—litmus paper turned violet. This, in my experience, is still the optimum pH of chicken soup.[2]

Whether this story is entirely true may be doubted,[3] but no matter—it tells a truth about young Piet Kolthoff. He loved chemistry and, very young, dove in deep.

Kolthoff was fortunate then to be both bookish and Dutch. The elite public high school in Almelo, his home town, demanded a commitment to studies that even today's over-stressed American high school achievers would find exhausting: plane and solid geometry, algebra, economics, world and national geography and history, 3 years of organic and inorganic chemistry, 4 years of English, 5 years of German, and 8 years of French. It suited him.

Completing the coursework marked a good start, but this was only a start. Then came two-and-a-half weeks of written tests, followed by oral exams. "For the oral exams we submitted a lists of 20 books ... we had been required to read in English, German, and French (not to speak of Dutch.) The two examiners could choose any one or more of these books and in addition to giving a summary of the contents by the candidate they would interrupt with questions." The language exam consisted of questions about the same books but in the language of the book. Fifty-five students started in Kolthoff's class; he and 10 others made it through.

This achievement still did not qualify him for a regular Dutch university, because he lacked knowledge of the Greek and Latin languages. So, in 1911, he enrolled in the pharmacy program at the University of Utrecht. He chose it so that he could get in, because the pharmacy program did not differ much from the chemistry program, and to study under Nicolaas Schoorl, professor of pharmaceutical and analytical chemistry.

As a second-year student at Utrecht, Kolthoff had come upon a book by Wilhelm Ostwald, *The Scientific Foundations of Analytical Chemistry*—and what youth could resist this title? There, he read that analytical chemistry, "the art of recognising different substances and determining their constituents," had been one of the "supreme importance" for centuries. But—and this is what most arrested Kolthoff—"analytical chemistry is content with fashions of theory which have long been discarded elsewhere and sees no harm in presenting its results in a shape which has really been antiquated for the last half-century."[4]

Young Kolthoff on holiday in Europe. Department of Chemistry, University of Minnesota—Twin Cities.

There it was: opportunity. A "supremely important" branch of chemistry had fallen far behind the rest of the field. Here lay an opening for an ambitious young man. At Utrecht, Schoorl and Kolthoff shared a wavelength. "Confining myself only to analytical chemistry, we spent one year in carrying out gravimetric and volumetric analysis, one year in organic analysis … one and a half years in pharmaceutical analysis, one quarter in food and water analysis, one quarter in toxicology, and one quarter in clinical analysis." For the rest of his long life, Piet Kolthoff praised Wilhelm Ostwald and Nicolaas Schoorl.

In 1915, at the age of 21 years, Kolthoff published his first paper, "Phosphoric Acid as a Mono- and Dibasic Acid." He got his Ph.D. in analytical chemistry in 1918. Over the next 5 years, Kolthoff published his first book, on pH, in 1920 and more than 150 scientific papers, a number impressive to everyone, except, it seems, the grandees of Dutch academia. Employed by the University of Utrecht as a chemist, he was permitted to begin teaching only in 1923 and then as a *privat docent*, not a professor, for no additional pay.

But Kolthoff's reputation had slipped the noose of the Dutch academy, and in 1924, he received an invitation for a summer lecture tour in the United States. He accepted, though nervously—he had just 4 years of classroom English. The journey took him first to Toronto, then to Rochester, New York. There, on a free day, young colleagues responded to his request to see something "typically American" by taking him to a burlesque house. Though he could not understand what the performers said, "visually speaking, I did not miss anything of the show."

In New York City, after a lecture to some pharmacists, came "beer, beer, and more beer," followed by a solo walk through Central Park, enlivened by an attempted mugging. "I was scared stiff, but I had a fortunate inspiration … I had learned a simple trick to break someone's leg. So I started taking my jacket off and asked [the mugger] 'Do you know Jiu Jitsu?'" When the mugger hesitated, Kolthoff sprinted away. His New York stay also featured a homosexual come-on, a three-card monte scam, and "more crazy experiences … with pickpockets, crooks, etc, than I have ever had since."

The most important part of the visit was still to come. Next, he went to Ohio State University, then Northwestern, and then University of Michigan, where he lectured at their departments of chemistry. These visits set the stage for Dean Lind's telegram 3 years later.[5]

We can guess what Lind had in mind when he extended his offer to Kolthoff. Samuel Colville Lind was a chemist himself, and a distinguished one. He had studied languages at Washington and Lee and chemistry at MIT and received his Ph.D. at the University of Leipzig. He studied for a year in Paris under Marie Curie, the discoverer of radium, and then more radiation studies in Vienna. After 7 years of teaching at the University of Michigan, he joined the U.S. Bureau of Mines as its radium expert. In 1926, he received the Nichols Medal, one of chemistry's highest honors. In the same year, Lind got faculty offers from both Michigan and Minnesota; he chose Minnesota, in part, for its (then) greater prestige.[6]

Samuel Colville Lind. University of Minnesota Archives, University of Minnesota—Twin Cities.

Lind had never met Kolthoff, but he had heard of him and read some of his papers. (Like Kolthoff, he was multilingual.) As his August 5 cablegram reveals, he had inquired about Kolthoff with H.R. Kruyt, a professor of physical chemistry at Utrecht who knew Kolthoff well; the two had spent many hours arguing. Kruyt had visited American universities and liked the close relations that he observed between universities and society.[7] Lind probably saw Kolthoff as a rising star; Kolthoff recalled decades later that Lind "was anxious to develop the scope of research at the University of Minnesota."[8]

Lind's cable came to Kolthoff "out of a blue sky," but it took him just 3 days to accept—why? There are hints in Kolthoff's writings. Despite his academic productivity, he was still earning so little—about $1200 a year—that he had to supplement his income writing for a Dutch newspaper. Kolthoff had found the

Dutch universities to be bound tightly by tradition, "concerned with the education of scholars and isolated from society." By contrast, he had seen, at Ohio State, Northwestern, and Michigan, big, young, well-funded universities, free from European traditions. "[I]n American universities education toward a bachelor's degree was directed to provide society with useful citizens." Lind's telegram also offered Kolthoff a good salary ($4500—about $60,000 in 2017 dollars) and the implied the promise of creating his own analytical chemistry program. This was an opportunity that Europe would never offer. Kolthoff arrived in Minneapolis 8 weeks later.[9]

## 1.3 THE PROFESSOR

He took an office in the chemistry building, later named Smith Hall, on the west side of Northrup Mall, and set about building an empire. He taught. He researched. He wrote. He traveled. He built networks. Today, we might apply the cliché word "workaholic," but the suggestion of a compulsion would get Kolthoff wrong. Work is what he wanted; it wasn't all he wanted, but it was what he wanted more than anything else. There were not enough hours in the day.

Kolthoff's teaching duties, as professor of analytical chemistry, and administrative duties, as section chairman, were relatively light. The main thing that had attracted him to Minnesota was the opportunity for research; he called the university his "El Dorado" for its laboratory facilities, talented graduate students, and the freedom that he enjoyed.

He put on the lab coat. Between 1928 and the end of 1941, Kolthoff published 232 scientific papers, in 19 journals and 3 languages, with 44 co-authors. The titles of these papers are mostly incomprehensible to anyone but chemists: "Effect of Dilution on the pH of Buffer Mixtures," from 1928, is one of the simple examples. They are mostly short, 5 to 15 pages (one, though, measured 94), variations on this theme: We went to the lab and tried this approach to solving a problem. Here is what we did, and these are the results. Studies like these, and countless others, published in journals, comprised the way in which chemists communicated with other chemists. Thus, little by little, Kolthoff pushed the growth of his discipline. And thus, little by little, Kolthoff broadened his own knowledge and reach. Here was Kolthoff relentlessly expanding the boundaries of analytical chemistry.[10]

The man and the hour had met. Twentieth-century science and industry demanded more and more precise and sophisticated chemical analysis. This could only be done by chemists with sufficient mastery of chemical theory to take on new problems. At Minnesota, Kolthoff established new standards for teaching graduate students, making sure that they were solidly grounded in physical chemistry theory—and in writing—for the clarity of description of the problems to be solved, the application of theory, and the succinctness of his experimentation.

Kolthoff with Department of Chemistry colleagues. Lind is at his right. Department of Chemistry, University of Minnesota—Twin Cities.

He published more than articles. In this period, 1928 through 1941, Kolthoff also published three books, starting with a translation of his 1927 *Die Massanlyse* (volumetric analysis); *Textbook of Quantitative Inorganic Analysis,* with his former graduate student Ernest B. Sandell; and *Polarography,* with another former graduate student James Lingane, in 1941.

The arithmetic: In his first 14 years at Minnesota, on average, Kolthoff published a scientific paper every 3 weeks, supervised a Ph.D. thesis every 6 months, and wrote or co-wrote a major scientific book every 5 years. He did not produce a flow, a river, or a flood but a tide of scholarship, all in the service of his big plan, the modernization and elevation of analytical chemistry as an academic discipline.[11]

Kolthoff excelled at attracting talent. The first was Ernest B. Sandell, later a co-author of *Polarography* and for many years a professor at Minnesota (and, with Kolthoff, creator of the Sandell–Kothoff test, still used for the measurement of iodine in human blood). James Lingane went on to a long career as a professor and department chair at Harvard. Herbert Laitinen, a Finn from Ottertail, Minnesota, taught chemistry—and trained his own cohort of graduate students—at Illinois, then the University of Florida, and served many years as editor of *Analytical Chemistry,* the first academic journal devoted to the subject. David Hume taught analytical chemistry for 33 years at MIT. Joseph Jordan taught analytical chemistry for 30 years at Penn State. Stanley Bruckenstein served more than 40 years as a professor at the State University of New York at Buffalo. Johannes Coetzee taught for many years at Pitt.

Every one of these teachers and scholars—and many others—extended the web of Kolthoff's influence, in space and in time, to later generations.[12]

In his campaign to modernize and elevate analytical chemistry, Kolthoff was a recruiter, drill sergeant, quartermaster, and chaplain: he found the young chemists, trained them, inspired them, and kept the resources flowing. He was the field marshal, master of strategy and tactics. He was a foot soldier too, working side by side with his recruits in the laboratory. Piet Kolthoff was not a man; he was an army.

By the end of 1941, Kolthoff had built an unshakeable reputation as a teacher, a scholar, and an innovator. The coming of World War II to the United States was about to take his career in a new direction.

## 1.4 WAR

From 1928 until 1940, Kolthoff took time every summer to visit his home town, Almelo. He had family there. His father, Mozes Kolthoff, died in 1936, but his mother still lived, along with his sister Elisabeth; her husband Jacob Wijler and their two daughters Marta and Rose; brother Abraham, and a host of uncles and aunts. As the 1930s went on, he felt the growing tensions of war in Western Europe. Almelo, in the northeast corner of The Netherlands, lay 10 miles from the German border.

The Kolthoffs were Jewish. After his last pre-war visit, in 1939, Kolthoff told a reporter, "I had a horrible feeling of premonition and there were tears in my eyes." Though Dutch Jews, among the most assimilated in Europe, had mostly remained relatively calm in this period, with few trying to leave the country, the Kolthoffs looked into it, moving money around in the hope of acquiring tourist visas to Uruguay. Nothing came of that. The Nazi army invaded Holland on May 9, 1940. The so-called Battle of The Netherlands lasted 7 days. Piet Kolthoff heard nothing from his family—Jews trapped in the Nazi occupation—until after Holland's liberation 5 years later.[13]

Mid-twentieth-century war ran on rubber: jeeps, trucks, armored personnel carriers, landing gear, and trucking transport on the home front. The British had controlled most of the world's rubber supply through its Southeast Asian colonies until they were conquered by Japan in the early 1940s. Synthetic rubber had been invented, but it played a small part in global supply. After Pearl Harbor, synthetic rubber became a crucial war material for the Allies, as Japan now controlled 90% of natural rubber resources. The United States began the war with less than a year's peacetime inventory. For the first year of the war, 1942, the country relied to an alarming degree on recycled rubber from a tire drive.[14]

Synthetic rubbers of limited usefulness had been developed in the early 1920s. German scientists working for I.G. Farben in the late 1920s and 1930s produced a tire-quality synthetic rubber, praised by Hitler and produced in modest quantities during the war. The I.G. Farben scientists had gotten U.S. patents in 1933, so their work was public knowledge.[15]

In late 1942, the U.S. Defense Rubber Committee selected Izaak Kolthoff as one of a dozen academic chemists to join chemists from Firestone, Goodyear, Goodrich, and, later, Phillips Petroleum, to take on a what was sometimes called a Second Manhattan Project—to find a recipe for synthetic rubber with two essential qualities: (1) durability comparable to that of natural rubber, and (2) the capability of mass production.

The accounts of the project's experiments provoke an image of chefs gathered around a giant cauldron, each with a big wooden paddle, stirring the mixture and arguing. There was a good deal of trial and error, because, according to Kolthoff, the participants so incompletely understood chemistry. Also, they were in a hurry.[16]

This was Kolthoff's first encounter with government-funded research, and he hated the bureaucratic nuisances—the meetings, the squabbles, and the red tape. On one occasion, he requisitioned a particular part, with precise specifications and available from just one vendor. When weeks passed without receiving it, he found that his request had been sent out for bid; the winning bidder produced an item only approximately the right size, that is, useless. He fumed.

He learned to bite his tongue while working with the oil industry representatives, who disliked collaborating with academics and tried to keep credit for themselves. "This lack of cooperation is disgusting," he wrote to the program's director.[17]

He also had to wrestle with the government over his researchers. Most of them were draft-age young men. Every time one of them got drafted, Kolthoff traveled to Washington, DC, to get the unlucky soldier undrafted, so that he could serve the war effort in the lab. Kolthoff claimed that he never lost one.[18]

Rubber research brought a lot of activity and money to the University of Minnesota. During the war years, Kolthoff's project employed up to 14 people (8 of whom, he wrote, were working 50 hours a week), including Kolthoff and his assistant. Edward J. Meehan, and paid three-fourths of Kolthoff's $5500 salary. In the years 1942–1946, the project brought in $231,000, about 3 million in 2017 dollars.[19]

Kolthoff's rubber work defies easy description. The manufacture of artificial rubber involves polymers, strings of molecules. If the strings are too short (not "poly" enough) or too long, the resulting rubber does not work well. So, one of the keys to the project was to find a recipe that halted the polymerization process at the right moment. In the words of the historian of this project, Peter Morris, "Kolthoff was drawn into the investigation of how the various parameters—temperature, soap, acidity, and modifier—influenced the rate or outcome of the polymerization." He and Herbert Laitinen, now at Illinois, developed new analytical techniques for the purpose.

By the end of war in 1945, the U.S. production of synthetic rubber had risen from 231 tons a year in 1941 to 920,000 tons in 1945. After the war, Kolthoff and his team moved past analysis and onto the big prize, improving the recipe. They played a part in making "cold rubber," a higher-quality product produced, literally, in the cold, about 20°F, that replaced natural rubber for tires forever. One of the two 1947 rubber patents that Kolthoff got provides a glimpse of him at work. The rubber recipe that the war group chose was called emulsion polymerization. To work well, it required the emulsion to be kept consistent, which was done by either shaking or stirring. Kolthoff, with colleague Charles Carr, devised a better way, known as ultrasonic waves.[20]

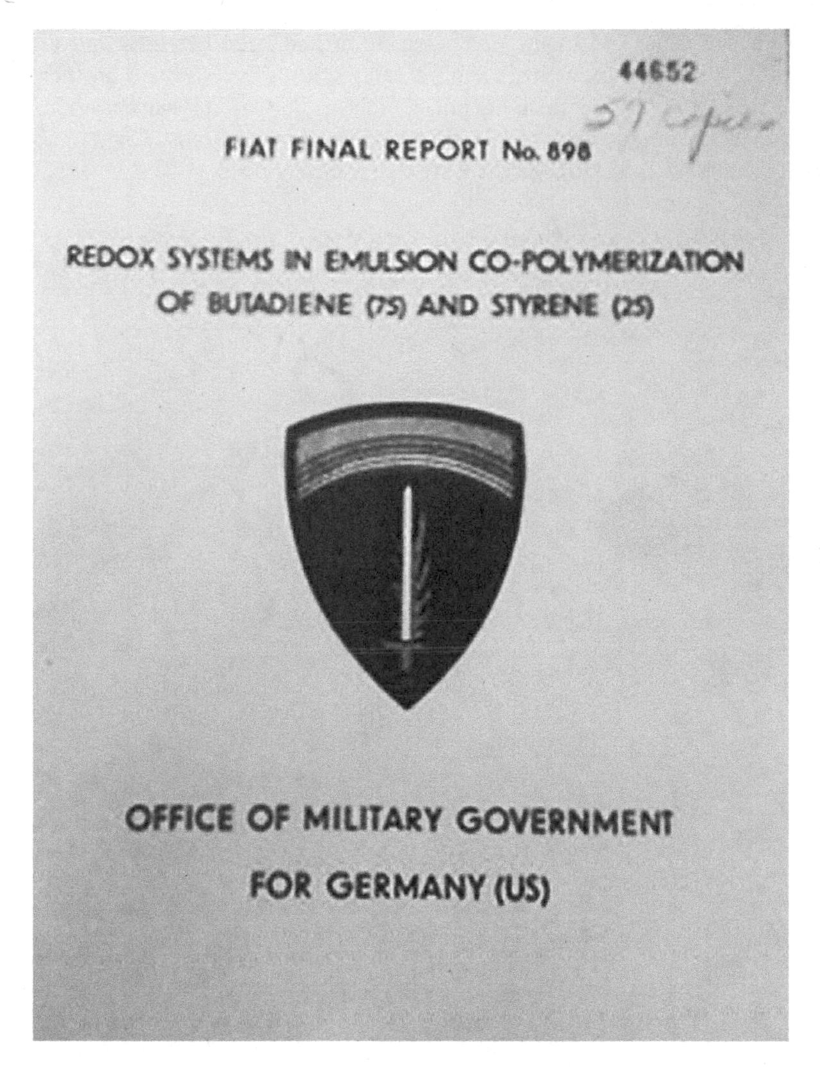
44652

FIAT FINAL REPORT No. 898

REDOX SYSTEMS IN EMULSION CO-POLYMERIZATION
OF BUTADIENE (75) AND STYRENE (25)

OFFICE OF MILITARY GOVERNMENT
FOR GERMANY (US)

Cover of a study Kolthoff produced for the U.S. government on a postwar visit to Germany. University of Minnesota Archives, University of Minnesota—Twin Cities.

Though Kolthoff sometimes downplayed the importance of his rubber work, he continued it for another 7 years, and near the end of his life, in a 1992 survey for *Who's Who in American Science*, he cited only one achievement: the "recipe for emulsion polymerization of rubber, 1942–1954."[21]

The Canadian Army liberated Almelo from the Nazis on April 5, 1945. Crushing news reached Kolthoff in fragments. Dutch Jews had been murdered at a higher rate than in any other occupied Western European country, almost 75%, compared with 45% in Belgium and 25% in France. Kolthoff's mother, her brother, and her sisters, all old by now, had been murdered at Vught, the Netherlands, and the Sobibor death camp in Poland in the spring of 1943. His two adult nieces, Marta and Rose, had gone into hiding (as did many Dutch Jews) but had been betrayed and killed on January 21, 1943, at Auschwitz. Their parents, Kolthoff's sister and brother-in-law, Elisabeth and Jacob Wijler, also hid and avoided capture, but when they learned the fate of their daughters, they took their own lives 6 weeks later. Only his brother, Abraham, also hidden, survived the war.[22]

Kolthoff's niece Marta, murdered in the Holocaust. University of Minnesota Archives, University of Minnesota—Twin Cities.

## 1.5 SUBVERSIVE

On April 5, 1951, Piet Kolthoff awoke to find his name in the morning *Minneapolis Tribune*, linked with film stars Judy Holliday and Mel Ferrer. The U.S. House of Representatives Committee on Un-American Activities (HUAC) had issued a report on prominent Americans associated with disloyal and subversive organizations. *The Tribune* headline featured Holliday and Ferrer; Kolthoff's name appeared lower in the text. The HUAC report named him six times.

Kolthoff was neither disloyal nor subversive, but he had done plenty to get himself noticed in that hypervigilant era. He had begun to take public political positions as early as 1930, when he became national president of the Association of Cosmopolitan Clubs, whose motto was "Above all nations, humanity." In 1931, he had written, "I frankly admit that I am a bad nationalist." This was innocent then, but in the Cold War, "internationalism" fell under suspicion as a synonym for "Communist sympathizer." Though—or, perhaps, because—Kolthoff had become a U.S. citizen in 1940 and a valued defense scientist in 1942, the FBI started investigating him in 1944. Agents quizzed his superiors at the University of Minnesota about his loyalty. We do not know whether they reported these contacts to Kolthoff.[23]

"U Scientist Flies to Russia." This headline from the July 23, 1945, *Minneapolis Tribune* could have been ominous—another turncoat? But it was merely descriptive. Within weeks of the end of World War II in Europe, the Soviet government invited Kolthoff and 15 other American scientists on a 17-day tour. We cannot know what the Soviets' motives may have been, but to Kolthoff the trip's only political element was the promotion of international scientific cooperation. After stops in Casablanca, Cairo, and Teheran, the Soviet tour began in Baku on the Caspian Sea, then Moscow; to a thousand-acre collective farm; then Leningrad; then back to Moscow, before undertaking a cross-Siberia air journey to Anchorage; and then home.

With the help of *The Tribune* staff writer Kenneth Scholes, Kolthoff published 10 articles about his journey, and in them, a government disloyalty hound could easily have found grounds for alarm. Kolthoff attended a dinner with Stalin and Marshal Zhukov; he described Stalin as displaying "great strength of character," praised the Soviet government for support of basic scientific research much better than that of the West, and noted that the escalator he rode in Moscow worked "at a much faster speed than escalators work in America." He seemed smitten by his U.S.-born Russian translator. In her version of the Soviet state, wages were fair, the necessities of life were amply provided, and freedom was abundant. "Natasha is enthusiastic about Soviet Russia, but I found her quite objective." Kolthoff's description of the collective farm was positive, though without overall judgment.

He wrote in praise of Russian theater and dance, the warmth of the Russian people, and the advancements of Russian science. The traitor-hunter would have noticed Kolthoff's pleasure at his encounters with chemist A. Frumkin and physicist James Frenkel; both were Jews who had taught in the Midwest, Frumkin at the University of Wisconsin and Frenkel at Minnesota, now living in the Soviet Union. In Moscow, he also ran into French scientist Frederic Joliot-Curie, a communist (also a Nobel Prize winner and son-in-law of Marie Curie) whose name featured prominently in the 1951 HUAC report.[24]

In 1948, Kolthoff was asked to join a "committee of 1,000" prominent citizens (Helen Keller was one of them) calling for the abolition of HUAC. The committee of 1,000 was called HUAC "betrayers of American ideals." Kolthoff joined gladly: "I am confident that the Un-American activities of the Un-American Activities Committee are of great concern to the majority of the American people."[25]

In 1949, Kolthoff visited Communist Yugoslavia and wrote about it for the Minneapolis *Star*. The country impressed him. "Yugoslavia was a backward and undeveloped country. The people had been exploited. Conditions were favorable for a social revolution. That revolution had brought about great social improvements …

Considering the poverty of the country and the recent developments of the socialized system, the results seem to be amazing."[26]

The 1951 HUAC report accused Kolthoff of association with six organizations that it called subversive: the Civil Rights Congress; Committee for Peaceful Alternatives to the Atlantic Pact [NATO]; Independent Citizens Committee for the Arts, Sciences and Professions; the Joint Anti-Fascist Refugee Committee; and Cultural and Scientific Conference for World Peace. So far as it went, the list was accurate—accurate but ham-fisted. These were all international peace organizations, mostly sprung up in reaction to the devastating recent European wars, the development of nuclear weapons, and the new threat of war between the West and the Communist Bloc. These organizations operated in public (that's how HUAC got the names) and were supported by hundreds of distinguished citizens. Besides Kolthoff and his red paramour Judy Holliday, née Judith Tuvim, a New York Jew, the report cited an all-star array of American artist and scientists, including:

- Performers: Singer and actor Paul Robeson (who really was a communist), actors Lee J. Cobb, Will Geer, Marlon Brando, Charles Chaplin, Jack Guilford, Gale Sondergaard (a Minnesotan), and Uta Hagen; dancer and choreographer Michael Kidd; and composers Leonard Bernstein and Aaron Copland
- Writers: Thomas Mann, Clifford Odets, Dashiell Hammett, Lillian Hellman, and Ring Lardner
- Scientists: Albert Einstein, Linus Pauling, and Karen Horney

Piet Kolthoff was not even the report's most-cited Minnesotan. That honor went to former Governor and U.S. Senator Elmer Benson. Kolthoff also came in behind a dangerous character named Charles Turck, president of Macalester College, St. Paul, then affiliated with the United Presbyterian Church, and recently elected president of the U.S. Council of Churches. The presidents of Augsburg College and Carleton College also made the list, along with Arthur Foote, pastor of Unity Unitarian Church in St. Paul.[27]

Kolthoff was more amused than threatened by the HUAC report, and he responded in typical Kolthoff fashion by writing directly to Congressman John Wood of Georgia, chair of the committee. Only the last of his three letters to Wood survives. On June 6, 1951, he wrote:

> I believe there is unanimous agreement that a war would result in a world catastrophe. From the Report of your Committee, one gets the uncomfortable impression that subscription to peace proposals and peaceful solutions makes a person a suspect of being a Communist sympathizer.

He then went on to correct the record by naming three *more* peace organizations that he belonged to, not mentioned in the report, and teased the Congressman for relying too much on what he read in a communist publication, *The Daily Worker*.[28]

Apparently, HUAC did not pursue Kolthoff further, but others did. In May 1953, the magazine *American Mercury* published a piece called "Communism and the Colleges." "In an endeavor to corrupt the teachers of youth," wrote J.B. Matthews, "the Kremlin has been remarkably successful, especially among professors at our colleges and universities." Kolthoff fell into the "fellow-traveler" category, not a Red, just a Red-abetter. "[T]hey all serve the purposes of Kremlin treachery."[29]

Kolthoff brushed it all off, but the fringe right wingers did not forget. As late as mid-1960s, an organization called Christian Research Inc. produced a pamphlet called *Communism at the U Today*. "HOW MANY PROFESSORS," it asked students, "... INDOCTRINATE STUDENTS WITH MARXISM-SOCIALISM ... RIGHT IN THE CLASSROOMS!??" Ranked first (despite being retired), "KOLTHOFF, Isaac M." with 35 "Communist front associations." (His chemical protégé, Cyrus Barnum, ranked fourth, with 12.)[30]

In fact, Piet Kolthoff had no use for communism or any ideology. Politically, he was a conventional liberal (he admired Hubert Humphrey) with a special interest in academic freedom and international cooperation among scholars, without political interference. Kolthoff recognized that science could be put to destructive use. In March 1945, he wrote to former Minnesota Governor Harold Stassen:

> [B]eware of the criminal nation that wishes to exploit science and its scientists for the ultimate destruction of mankind. Although the scientist should maintain complete freedom in his choice and pursuit of research, I feel that the progress and application of scientific research should be closely supervised by an international board.[31]

He belonged to United World Federalists, and in 1949, he had written that world government was the only hope for avoiding future wars.

Kolthoff was less idealistic than some of the intellectuals who signed the same calls for international peace and cooperation that he sometimes did. In 1948, he met in a small group with Henry Wallace, the most left-leaning man ever to hold the Office of Vice President, and about to run against President Truman in the fall election. He put the question to Wallace, "What assurances are we going to demand from Russia to protect our security?" and was disappointed for a lack of response. "I am not joining the Wallace group," he wrote a few days later, "until I have assurances that it is not only a matter of giving to but also taking from the Russians." He had written a similar opinion to Albert Einstein.

In 1950, he had been asked to join the World Congress of Defenders of Peace in a call for nuclear disarmament. "Do you really think that outlawing any weapon will contribute to peace? War is immoral, and under desperate circumstances, any nation will use any weapon for self-preservation." He suggested a different approach: "I would like an appeal to all nations that they submit to international control of elementary human rights."[32]

Kolthoff lecturing in Moscow, 1958. Department of Chemistry, University of Minnesota—Twin Cities.

The HUAC accusations did not hurt Kolthoff professionally, but they did sting. From the onset of World War II forward, he had been an enthusiastic advocate of defense research. In 1951, he had joined the Chemical Advisory Committee of the Air Research and Development Command. But he lost his security clearance in late 1952 and was forced to resign. He got it back in December 1953 but not before having been subjected twice to fingerprinting. This must have galled him; he had long been on record, opposing such measures.[33]

The accusations had no effect on Kolthoff's scientific productivity. The demands of wartime research slowed his rate of publication to just 18 papers, 1942–1945, but soon, he got back to full production. From 1946 through his retirement year, 1962, he published another 241 papers, about 14 a year. The rubber research pushed him into a new area of publication, colloidal chemistry. (Emulsions are one variety of colloids, and the cold rubber recipe that Kolthoff had helped to develop was called "emulsion polymerization." He had first studied colloids with Schoorl in Utrecht.) He also returned to cancer research, an area he had first explored in 1939 (blood analysis), and published articles in *The Lancet* and the *Journal of the National Cancer Institute*, two publications normally outside of his already wide scope.[34]

In addition to his technical publications, Kolthoff also wrote and spoke regularly about the history of analytical chemistry and the proper way to educate scientists. Above all, he advocated liberty of thought and urged his own students not to accept received wisdom or even his own wisdom if they disagreed. He rejected requests from industry to train his students for industry. The qualities that he prized were "imagination, originality, initiative, judgment, and curiosity."[35]

Still not sufficiently busy, in this period, Kolthoff also worked part-time as an editor for New York publisher Interscience and its chemical analysis series of monographs. This had started in 1942, slowly, and then accelerated in 1950s and 1960s. Kolthoff may have contributed to as many as 50 of these books. He also became a major shareholder in the company.[36]

His own books kept coming too, six monographs based on graduate students' Ph.D. theses (with Kolthoff as co-author); *Emulsion Polymerization* (with three co-authors), a description of his rubber work; and another book on pH.

Around 1954, he began to think of something bigger than he had ever done, an encyclopedic treatment of his field. The first volume of *Treatise on Analytical Chemistry* came out 5 years later. One sentence in the foreword sums up the size of his ambition:

> The aims and objectives of this Treatise are to present a concise, critical, comprehensive, and systematic, but not exhaustive, treatment of all aspects of classical and modern analytical chemistry. Can we go over that again? "Concise, critical, comprehensive, and systematic … all aspects of classical and modern analytical chemistry." Whew!

In the end, it is not clear that he achieved concision. Part 1 came out in 1959—10 volumes; though, to be fair, all are slim, just 400–500 pages each. Part 2, 16 volumes plus index, came out in 1961, and Part 3, 4 volumes, was issued over a 10-year period, the last in 1977, 15 years after Kolthoff had "retired."

The reader is invited to admire not only the ambition of the task but also the scope. Over 18 years, the *Treatise* engaged 295 authors, from 60 universities, 55 private companies, 18 government departments, 12 research institutes, and 8 countries; contained 245 articles; and consumed over 17,000 pages. The *Treatise on Analytical Chemistry* made a fitting culmination to a career that was still not over.[37]

A non-chemist who paged through the *Treatise* would discover a vast and unknown world, one where everyone understands the word "titration," where careers are made observing vanadium, and where industries turn on the precise arrangement of hydrocarbon molecules. It's a realm where the impenetrable (to almost everyone) is common currency. At the same time, some article titles show where any citizen can understand the everyday application of analytical chemistry: "Testing of Consumer Products," and "Potable and Sanitary Water." The people who identified the hazards in the Flint, Michigan, water supply in 2016 were practicing analytical chemistry.

The University of Minnesota required Kolthoff to retire in 1962, as he neared age 70 years. Nothing else changed much. He continued to live on campus, as he had done since 1927; he had a room in the Campus Club on an upper floor of Coffman Memorial Union. He continued to experiment, write, and publish. The chemistry department allowed him lab privileges and space for up to three post-doctoral researchers. Kolthoff took that as a starting offer: for the next decade, he employed at least 10 every year.[38]

In retirement, he published 126 papers, many of these with his devoted laboratory colleague Dr. Miran K. Chantooni, Jr., who stayed with him for over 30 years. In research there was no looking back; Kolthoff continued to keep up with the literature in his field and explore new areas. A prominent chemist noted that if Kolthoff had started over in 1962, as a fledgling professor, his post-retirement record of research and publication would have qualified him for tenure at almost any university.[39]

He traveled more, to Florida and the Cayman Islands for fishing, and to Israel, where he had relatives and had taken a great interest in promoting the study of chemistry at Israeli universities. He played bridge; he received bushels of mail, especially on the occasions of his 80th, 90th, and 95th birthdays. He often worked outside his room at the Campus Club and had an informal invitation to join any buffet line that appeared nearby.

By 1979, he complained that he could only sustain concentration for a few hours a day and took leave of his editing duties at Wiley-Interscience. The sports injuries that Kolthoff had suffered over the years—skiing and horseback riding—took a predictable toll on his mobility. In 1981, when he had reached 85 years of age, the university arranged to move him a few dozen yards to the west, into Comstock Hall, where he lived at the university's expense. Among his many distinctions was that he was the only male professor at the university to live in a women's dormitory.

Only at the very end of his life, in his late 90s, did Piet Kolthoff finally leave the campus. He moved to the high-rise apartments at 740 East Mississippi River Boulevard, St. Paul, and, ultimately bedridden, to Bethesda Rehabilitation Center, where he died on March 4, 1993.[40]

## 1.6 THE FATHER OF MODERN ANALYTICAL CHEMISTRY

More than 21 years after Piet Kolthoff's death, in a ceremony held at Smith Hall, the American Chemical Society designated the contributions of Izaak Kolthoff to modern analytical chemistry as a National Historic Chemical Landmark. The Chemical Landmark was Kolthoff's greatest award but far, far from the first. In the last half of his life, receiving awards became a recurring thing. He got plenty of opportunities to tell his "optimal pH for chicken soup" story.

In 1949, Kolthoff received the Nichols Medal, for original research in chemistry, from the New York division of the American Chemical Society. Previous and future Nichols laureates included Linus Pauling, Irving Langmuir, Glenn Seaborg, Robert Woodward, and Vincent du Vigneaud, all Nobelists in chemistry, and James B. Conant, chemist, U.S. Ambassador to West Germany, and president of Harvard University.[41]

The next year, Kolthoff received the Fischer Award in Analytical Chemistry from the American Chemical Society. The award speech noted that "more research papers have come from the prolific pen of this year's recipient than from any other living analytical chemist" and that he had also educated more Ph.D. students in analytical chemistry than anyone else. He was just the third to win it, and his influence affected this award for years to come. His former graduate students James Lingane, Herbert Laitinen, and David Hume won it in 1958, 1961, and 1963, respectively–four U of M students or professors in 14 years. Kolthoff's co-editor of *Treatise on Analytical Chemistry*, Philip Elving, won it in 1960.[42]

In 1964, the Chicago Section of the American Chemical Society gave Kolthoff its 53rd annual Willard Gibbs Award. This one put him in a company almost as distinguished as the claque of traitors identified with Kolthoff in the HUAC report of 1951: 17 then and future Gibbs medalists won Nobel Prizes in chemistry, including Linus Pauling, again, and Marie Curie.

Chemists will understand the specifics of Kolthoff's Gibbs:

> In recognition of his unique influence on, and his manifold contributions to, the understanding, practice, and teaching of analytical chemistry, as exemplified by his fundamental studies of classical titrimetry; indicators; pH and buffer solutions; coprecipitation and the aging of precipitates; polarography; the kinetics and mechanisms of emulsion polymerization; and potentiometric, conductiometric and amperometric titrations.

Anyone can understand the award's general purpose: "To publicly recognize eminent chemists who, through years of application and devotion, have brought to the world developments that enable everyone to live more comfortably and to understand this world better." This fit Kolthoff perfectly, especially the phrase, "years of application and devotion."

In 1967, the Academy of Pharmaceutical Sciences created a new award, the Kolthoff Gold Medal Award in Analytical Chemistry "to recognize long-term significant research advancing the science of pharmaceutical analysis." The medal's first recipient? Izaak Kolthoff. In his acceptance speech, he told the chicken soup story.

In 1972, the University of Minnesota used the last sliver of buildable land on Northrup Mall, the center of its main campus, to put up a new chemistry center, Kolthoff Hall. At the dedication ceremonies—and elsewhere again and again—people honored Kolthoff with the title "father of modern analytical chemistry."

With University of Minnesota President C. Peter Magrath at the dedication of Kolthoff Hall. University of Minnesota Archives, University of Minnesota—Twin Cities.

The key word in that phrase is "modern." As Wilhelm Ostwald wrote late in the nineteenth century, analytical chemistry then was already hundreds of years old, but its scientific bases were antiquated, obsolete. The "science" was not much more than a compilation of lab techniques.

Kolthoff's great contribution was to reconnect laboratory practice with theory. The word theory is often misunderstood. In common speech, it often means just a guess: "That's MY theory." In science, theory is not guesswork but an organized body of knowledge that has both analytical and predictive powers. The theory of biological evolution, for example, both explains the fossil record and predicts the mutation of, say, the Vika virus.

Kolthoff did not produce new theories, like Darwin and Einstein did. Instead, he worked to integrate the already-existing theories in chemistry, especially physical chemistry (the structure and nature of the atom), into chemical analysis. This was, in a way, a harder task than theorizing. Darwin and Einstein caught people's attention with their spectacular notions. All Kolthoff undertook to do was modernize his science from the ground up, through teaching, writing, publishing, and persuading. The Gibbs Award got it exactly: "years of application and devotion."

His own motto was "theory guides, experiment decides." We might think of experiments leading to the formation of theory, but Kolthoff put it the other way. Analytical chemists equipped with understanding of modern chemical theory had the tools not just to conduct chemical analysis but also to understand the "why" of their results. With the understanding of "why" came also the ability to imagine new tools and techniques. Einstein's theory of general relativity, 1903, provoked the imagination of the "God particle." In 2016, experiments confirmed its existence. Theory guides, experiment decides.

When he began at Minnesota, analytical chemistry was a technical specialty. By the time Kolthoff retired, the field had achieved full parity with the other branches of chemistry. Two scientific journals, *Analytical Chemistry* and *Talanta*, were devoted to it. All major universities employed professors of analytical chemistry, many of them trained by Kolthoff and his disciples. An analytical chemist, and friend of Kolthoff, Jaroslav Heyrovsky, won the Nobel Prize in 1959. It was estimated in 1964 that 1100 analytical chemists around the world, in universities and industry, traced their academic pedigrees to Piet Kolthoff, the father, grandfather, and now, in 2016, great-grandfather of modern analytical chemistry.

One of his graduate students, David Hume (who went on to teach analytical chemistry at MIT), tried to sum up Kolthoff's achievements:

> More than any other person, he transformed analytical chemistry from a rather unimaginative, conservative art to a dynamic, innovative science. He swept the profession along with him … So widely is his style of approach now used that … we tend to take it for granted. And we forget, if we ever knew, what the source was.[43]

In 1994, the American Chemical Society published a book, *Landmarks in Analytical Chemistry*, a compilation of exemplary papers in the field. For the period 1930s–1940s, the editors chose one Kolthoff paper among the 10 exemplars. It bore a typically Kolthoffian title: "Polarographic Determination of Manganese as Tridihydrogen Pyrophosphatomanganiate." Try saying that three times fast.

Alan Bond of LaTrobe University in Australia wrote in appreciation:

> This paper represents one of the earliest demonstrations of how an important problem was elegantly solved by using a carefully integrated combination of Classical chemical-based methods, an instrument method [the polarograph], and relevant and recent theory.

An elegant combination of classic methods, new methods, and theory—what could be better in any profession? "Fifty years later," Bond continued, "this analytical protocol represents the standard expected in analytical chemistry."[44]

Kolthoff's posthumous Chemical Landmark award was his greatest. The National Historical Chemical Landmarks program goes back to 1993 and stands out among scientific awards for its range, from the practical to the erudite. The first award went to Bakelite, a pioneering plastic now prized by collectors. With his award, Kolthoff joined a roster that includes pioneering chemists Joseph Priestley and Antoine Lavoisier, the American agronomist George Washington Carver, water-based paint, Scotch tape, Tide detergent, the legacy of Rachel Carson's *The Silent Spring*, and, in a kind of double-up for Kolthoff, the U.S. Synthetic Rubber Program, in which Kolthoff played so big a part.

Thomas Edison and Piet Kolthoff, alike in their "years of applications and devotion," were added as "chemical landmarks" in the same year.[45] One of the reasons why Kolthoff left Europe for Minnesota in 1927 was the lure of a university that took an interest not just in pure research (which Kolthoff supported) but also in bringing practical benefits to society. For Kolthoff to be placed by chemists who followed him in a pantheon of great theorists (Lavoisier and Priestley), great practical innovators (Edison and Carver), people of conscience (Rachel Carson), and things that made people's lives better (synthetic rubber, water-based paint), would have pleased him to no end.

Program for Chemical Landmark Award. University of Minnesota Archives, University of Minnesota—Twin Cities.

## 1.7 EPILOG

Kolthoff died in 1993. His last paper, written with M.K. Chantooni, was published after his death. The subject was crown ethers, a kind of designer molecule, which for Kolthoff, in his 90s, was a new area of inquiry. Just as there had not been enough hours in the day to do all he wanted to do, there were not enough years in his life. "I could have accomplished a lot more." he once said, "if I had worked harder."[46]

## ACKNOWLEDGMENTS

I am indebted to several people and institutions for the privilege of writing about Izaak Kolthoff. This project began with a chance meeting with the University of Minnesota Department of Chemistry Chair Professor William Tolman at the Minnesota State Fair in 2015. That conversation introduced me to Kolthoff.

Kolthoff's papers are in the University of Minnesota Archives; my thanks to the people there for their professionalism and help.

Professor Peter Carr made the Kolthoff Notebooks available to me, which were compiled by Professor Kolthoff's longtime secretary, late Christa Elguther. Without her preservation and organization of Kolthoff's papers, much of his story would have been lost. According to Pete Carr, Ms. Elguther provided a "serious multiplier" of Kolthoff's accomplishments, especially in his later years.

Pete Carr and Professor Mark Vitha of Drake University were kind enough to read this manuscript and offer critically helpful corrections and suggestions.

I could not have traced the fates of Kolthoff's family members killed in the Holocaust without the help of Ariane Zwiers of the Jewish Cultural Center, Almelo, The Netherlands. It was she who identified the photo of Kolthoff's niece, Marta.

## END NOTES

1 Cablegram, S.C. Lind to I.M. Kolthoff, August 5, 1927; Cablegram, I.M. Kolthoff to S.C. Lind, August 8, 1927. I.M. Kolthoff Papers, University of Minnesota Archives [hereafter, Kolthoff Papers.]

2 I.M. Kolthoff. "Autobiography." Kolthoff Notebooks, University of Minnesota Department of Chemistry [hereafter, Kolthoff Notebooks.]

3 It does not hold up. No cook in her own kitchen would long confuse the powdery baking soda with the chunky, crystalline table salt. And no cook would put "several large spoonfuls" of salt into a soup, unless it were in a vat prepared for an army.

4 Ostwald, W. *The Scientific Foundations of Analytical Chemistry*. London, UK: MacMillan & Co., 1900, preface to the first German translation, vii–viii.

5 "Autobiography." Kolthoff Notebooks. According to Peter Carr, who knew Kolthoff well, this story is "pure B.S."

6 Laidler, Keith. "Samuel Colville Lind, 1879–1965." *Biographical Memoirs* v. 74. Washington, DC: National Academy Press, 1998.

7 Van Berkel, K., and Albert Van Helden, L.C. Palm, eds. *History of Science in the Netherlands: Survey, Themes, and Reference*. Leiden, Boston, Koln: Brill, 1999, pp. 182, 209, 508.

8 Brasted, Robert C. "Interview with I.M. Kolthoff." *Journal of Chemical Education* 50, October 1973, p. 663.

[9] Kolthoff, I.M. "Life as an Analytical Chemist," typescript 1981, Kolthoff Papers.

[10] "Publications of I.M. Kolthoff." *Talanta* 11, no 2 (February 1964), pp. 351–385.

[11] Author's compilation of Kolthoff publications in the University of Minnesota libraries.

[12] The web site Chemistry Tree provides an excellent, though not complete, listing of chemists tracing their academic lineage to Kolthoff.

[13] *Minneapolis Times Tribune*, May 27, 1940, p. 19; Moore, Bob. *Victims and Survivors, The Nazi Persecution of the Jews in the Netherlands 1940–1945*. New York: St. Martin's Press, 1997. Letter, I.M. Kolthoff to Guarantee Trust Co., of New York, attempting to borrow $10,000 against the assets of Louis Wijler and J. Wiljer for nineteen tourist visas for Uruguay, for family members. Folder labeled Correspondence 1928–1944, Kolthoff Papers.

[14] Rockoff, Hugh. "Keep On Scrapping: The Salvage Drives of World War II." Cambridge, MA: National Bureau of Economic Research Working Paper No. 13418, September 2007, pp. 24–39.

[15] Herbert, Victor, and Attilio Bisio. *Synthetic Rubber, A Project That Had to Succeed.* Westmont, CT: Greenwood Press, 1985, pp. 25–33.

[16] Morris, Peter J. *The American Synthetic Rubber Research Program*. Philadelphia, PA: University of Pennsylvania, 1989, pp. 13–38. Still working on rubber in 1956, Kolthoff applied for a National Science Foundation grant to further study emulsion polymerization. He wrote, "[I]t must be admitted that our understanding of the kinetics and mechanism of any emulsion polymerization system is most inadequate. The recipes now used by the American Rubber Industry [and largely produced by Kolthoff and his colleagues] have been developed almost entirely on an empirical basis … ." Grant application, folder marked "Grant Papers," Kolthoff Papers.

[17] Kolthoff to Dr. C.S. Fuller, November 8, 1943. Kolthoff Papers, folder labeled Correspondence 1926–1944.

[18] Brasted interview.

[19] Department of Chemistry records, U of M Archives, folder marked "Dr. Kolthoff's Rubber Research." Through June, 1952, the University received $631,214. Only the Universities of Illinois, and Chicago, and Case Institute in Cleveland, got more. Herbert and Bisio, *Synthetic Rubber*, 149. The University of Illinois program was run by Kolthoff's former graduate student, Herbert Laitinen.

[20] Bovey, F.A., I.M. Kolthoff, A.I. Medalia, and E.J. Meehan. *Emulsion Polymerization*. New York: Interscience Publishers, 1955. Herbert and Bisio, *Synthetic Rubber*. Kolthoff and Carr patent.

[21] Morris, Peter T. *The American Synthetic Rubber Research Program*. Philadelphia, PA, University of Pennsylvania, 1989; American Chemical Society. "United States Synthetic Rubber Program, 1939–1945" (pamphlet produced for dedication of a chemical landmark in Akron, August 29, 1998.) Washington, DC: American Chemical Society, 1998. Kolthoff is not mentioned in the brochure, but this sentence—"University laboratories developed better analytical methods to achieve better quality control and performed fundamental research on GR-S polymerization and the chemical structure of rubber."—probably refers to the work done at Minnesota. Questionnaire, Who's Who in Science and Engineering, 1992–1993, Kolthoff Papers.

[22] Van der Boom, Bart. "Ordinary Dutchmen and the Holocaust: A Summary of Findings," in *The Persecution of the Jews in the Netherlands*. Amsterdam: University of Amsterdam Press, 2012 [page?] Information about the fates of Kolthoff's relatives comes from a variety of sources, the Jewish Historical Museum of Amsterdam and Culemborg (where his mother's family came from), and the Yad Vashem Central Database of Shoah victims. Personal communications from Ariane Zwiers, Jewish

Cultural Quarter, Almelo. https://www.joodsmonument.nl/en/page/479780/jacob-samuel-wijler-and-his-family, accessed November 14, 2016. DutchJewry.org, http://www.dutchjewry.org/genealogy/duparc/2014.shtml, accessed November 19, 2016 (Jacob Wijler family); http://www.dutchjewry.org/genealogy/duparc/2014.shtml (Rosette Wijzenbeek's siblings).

23 Kolthoff, Isaac [sic] M. "Nationalism and Internationalism." *The Cosmopolitan Student* XXI(4), May 1931, p. 1. FBI file on Kolthoff, obtained by the author through a Freedom of Information Act request.

24 *Minneapolis Tribune*, July 23, 1945, p. 1; July 24, 1945, p. 1; July 25, 1945, p. 2; July 26, 1945, p. 2; July 27, 1945, p. 2; July 28, 1945, p. 2; July 29, 1945, p. 2; July 30, 1945, p. 2; July 31, 1945, p. 2; August 1, 1945, p. 3. Goldsmith, Maurice. *Frederic Joliot-Curie: A Biography*. London, UK: Lawrence and Wishart, 1976.

25 IMK to Harlow Shapley, January 12, 1948, Kolthoff Papers, Box 1, Correspondence 1948–1949.

26 *Minneapolis Star*, October 19, 1949, p. 19.

27 Committee on Un-American Activities, U.S. House of Representatives, "Report on Communist 'Peace' Offensive, A Campaign to Disarm and Defeat the United States." Washington D.C., April 1, 1951.

28 IMK to Congressman John Wood, June 6, 1961. Kolthoff Papers, Box 1, Correspondence 1960–1962.

29 Matthews, J.B. "Communism and the Colleges." *The American Mercury*, May 1953, pp. 111–146. Only three other Minnesotans were named: Charles Turck, again, and University of Minnesota professors Henry Burr Steinbach (biology) and Joseph Weinberg (physics.) Neither Turck nor Steinbach seem to have been hurt by such accusations—Steinbach had a long career at the University of Chicago and Woods Hole National Laboratory. *New York Times*, December 24, 1981. Weinberg, a member of the Manhattan Project, was dismissed by Minnesota in 1953, over suspicions about his loyalty, and returned to the academy only in 1957, at Western Reserve University. He then went on to a long career at the University of Syracuse. http://archives.syr.edu/collections/fac_staff/sua_weinberg_j.htm.

30 Kolthoff Notebook no. 16.

31 IMK to Harold Stassen, March 9, 1945. Kolthoff Papers, Box 1, Correspondence 1945–1947.

32 IMK to Frederic Joliot-Curie, May 10, 1950, Kolthoff Papers, Box 1, Correspondence 1950–1951. Prof. Joliot-Curie was a scientist whom Kolthoff knew, and an enthusiastic Communist and supporter of the Soviet Union.

33 Amos Horney of Air Research Group to IMK, [date?]; Paul Flory of Cornell University to IMK November 22, 1952; IMK to A.H. Blatt, Chairman of Chemistry Advisory Committee, Air Research and Development Command, August 31, 1953; Irene Backstroh, University of Minnesota Security Office, to IMK, January 6, 1954; IMK to Members of Congress, March 6, 1950, Kolthoff Papers, Box 1, Correspondence 1950–1952 and 1953–1954.

34 Stricks, W; Kolthoff, IM; Bush, DG; Kuroda, PK. "Experimental Study of the Polarographic Cancer Test of the Sulfhydryl Titration for the Differentiation Between Normal, Cancerous, and Other Pathological Sera." *Journal Lancet* 73(8), August 7, 1953, 328–335; "Nitrogen, sulfhydryl, and disulfide contents in normal and cancer sera and in their albumin and globulin fractions. Action sulfite on disulfide groups in normal and cancer sera (with W. Stricks, Loes Morren, and A. Heyndrickx). *Journal of the National Cancer Institute,* 1954, 1233–1245.

35 Kolthoff, I.M. "Analytical Chemistry as a Technique and as a Science." *Chemical and Engineering News,* 28(4), August 21, 1950, pp. 2882–2886.

[36] Personal communication with Prof. Mark Vitha of Drake University, current (2017) editor of the *Chemical Analysis* series.

[37] Author's compilation from the first edition of the *Treatise*, with the assistance of Prof. Mark Vitha of Drake University. Oddly, perhaps, the University of Minnesota libraries do not own a complete set. Prof. Vitha, who got his PhD at Minnesota, owns a set previously owned by I.M. Kolthoff.

[38] Ray Archer, University of Minnesota Department of Insurance and Retirement to IMK, February 20, 1961 (retirement), Kolthoff Papers, Box 1, Correspondence 1960–1961; Stuart Fenton, Chemistry Department Chairman, to IMK, June 6, 1962 (lab space), Kolthoff Papers, Box 1, Correspondence 1962–1963.

[39] Kolthoff Papers, Box 2, Kolthoff's Bibliography (126 papers); Box 3, List of Postdoctoral Fellows 1951–1980. He employed over forty post-doc researchers after his retirement.

[40] *Minnesota Daily,* March 8, 1993.

[41] Nichols Award web site: https://www.acs.org/content/acs/en/funding-and-awards/awards/acs-local-section-awards/new-york-local-section.html.

[42] Typescript of award remarks, Kolthoff Papers, Box 1, Correspondence 1947–1949. Fischer web site. Kolthoff's chemical grandson (former graduate student of Kolthoff's former graduate student, Joseph Jordan) and Professor of Chemistry at Minnesota, Peter Carr, also won this award in 2009.) Carr believes Kolthoff's greatest accomplishment lay in the education of his graduates students.

[43] Typescript, "Professor David M. Hume in an improvised talk discussed Kolthoff's contributions to science," upon the dedication of Kolthoff Hall. Kolthoff Notebook 16.

[44] Warner, Mary, and Louise Voress, Grace K. Lee, Felicia Wach, and Deborah Noble, eds. *Milestones in Analytical Chemistry.* Washington, DC: American Chemical Society, 1994, p. 26.

[45] Directory of National Historic Chemical Landmarks, https://www.acs.org/content/acs/en/education/whatischemistry/landmarks/landmarksdirectory.html, accessed November 1, 2016.

[46] Coetzee, Johannes F. "Izaak Maurits Kolthoff 1894–1993," in *Biographical Memoirs, Vol 77.* Washington DC: National Academy Press, 1999, p. 17.

# 2 Recent Advances in Two-Dimensional Liquid Chromatography for the Characterization of Monoclonal Antibodies and Other Therapeutic Proteins

*Dwight R. Stoll, Kelly Zhang, Gregory O. Staples, and Alain Beck*

**CONTENTS**

## 2.1 OVERVIEW

Characterization and analysis of monoclonal antibodies (mAbs) and related therapeutic proteins are very challenging from an analytical point of view. This is mainly because of the large size of these molecules and the importance (e.g., effects on stability, toxicity, and efficacy) of small differences (e.g., deamidation) between molecules that are otherwise chemically identical and because of the inherent heterogeneity of these materials that reflects the biosynthetic processes that produce them. Two-dimensional liquid chromatography (2D-LC) is an analytical tool that has a lot to offer to scientists looking for ways to effectively and efficiently address the challenges associated with analysis of mAbs and related proteins. In 2016, we published a first-of-its-kind review article focused on the recent developments in 2D-LC for the analysis of mAbs and related materials [1]. Since that time, there has been a good deal of research published on the development of next-generation therapeutic proteins, on innovations in 2D-LC as an analytical platform, and specifically on the application of 2D-LC to analyze mAbs and related materials. Our primary aim in this chapter is to provide an update on what has happened in these fields since the publishing of our 2016 review. As such, it is not intended to be an exhaustive review of any of these research areas. Table 2.2 in Section 2.4.1 contains references relevant to each area addressed in this review. This table can serve as a resource to readers interested in quickly learning what has been done in those areas, including both work cited in the 2016 review and work that has been published since then. The accompanying text, however, is focused mainly on new work from the last 2 years or the most recent work in areas where little has been published during that time.

## 2.2 INTRODUCTION TO CHARACTERIZATION OF ANTIBODIES AND OTHER THERAPEUTIC PROTEINS

MAbs and related products are the fastest-growing class of therapeutics for human use. More than 80 isotype G immunoglobulins (IgGs) and IgG derivatives have been approved for use in various indications such as cancer and inflammatory diseases [2] and over 50 more are currently being investigated in late clinical trials [3]. The IgG-based pharmaceuticals approved by the U.S. Food and Drug Administration and the European Medicine Agency include "naked" antibodies, radio-immunoconjugates, antibody–drug conjugates (ADCs), bispecific antibodies (BsAbs), antigen-binding (Fab) fragments, Fc-fusion proteins or peptides, and immunocytokines [4]. In addition, marketing authorizations for a number of biosimilar antibodies have recently been granted in both Europe and the United States [5].

In the past decade, more than 1000 papers have been published on the analytical and structural characterization of mAbs, and the rate of publication has increased significantly during the last 5 years. Multiple and complementary LC, electrophoresis, and mass spectrometry (MS) methods are used at all stages of mAb discovery and preclinical and clinical development [6]. These analytical techniques are helpful for the selection of the best antibody-producing clone generating the desired glycoprofile [7] and for full structural characterization of research leads and clinical candidates [8]. A combination of LC, electrophoresis, and MS is also used for the identification of "hot spots," which may be deleterious for drug product stability, for pharmacokinetics (PKs), and for pharmacology properties [9]. Consequently, these characterization methods are essential for biochemical, biophysical, and potency assays; in support of formulation, process scale-up, and transfer; and in defining critical quality attributes (CQAs) in a quality-by-design (QbD) approach. Importantly, the use of MS methods early in the research and development process also helps one to optimize the structure of next-generation mAbs from a pharmaceutical standpoint, allowing the development of candidates with reduced chemistry, manufacturing, and control (CMC) liabilities and better drug-like properties.

A variety of separation techniques based on both LC and electrophoresis hyphenated to MS are used for antibody characterization and homogeneity assessment. These orthogonal analytical methods aim at separating the main antibody isoforms from other variants and process-related impurities such as host cell proteins (HCPs).

## 2.3 INTRODUCTION TO TWO-DIMENSIONAL LIQUID CHROMATOGRAPHY

In this section, we briefly review the main principles of 2D-LC and comment on some aspects of 2D separations that are especially important in the analysis of mAbs and related materials. Readers interested in more detailed discussions of the principles of the technique and recent methodological developments are encouraged to consult several recent reviews [10–12] and a book chapter [13].

The attractiveness of 2D separations in general is illustrated in Figure 2.1 Here, we suppose that we have a sample composed of molecules varying in both size and charge (in this case, three levels of size and three levels of charge).

Considering this hypothetical sample, separating the mixture using conventional one-dimensional (1D) chromatography with the help of a column that is selective for size, we end up with partial resolution of the sample, with peaks composed of co-eluting analytes that vary by charge (Figure 2.1a). Similarly, if we separate the mixture using a single column that is selective for charge, we end up with partial resolution of the sample, with peaks composed of co-eluting analytes that vary by size (Figure 2.1b). The central concept of 2D separation is that we combine the two complementary columns in such a way that analytes co-eluting from the first-dimension ($^{1}$D) column are resolved by the second-dimension ($^{2}$D) column after fractions of effluent from the $^{1}$D column are transferred (i.e., injected) from the outlet of the $^{1}$D column to the inlet of the $^{2}$D column. In the best case, this then results in the complete resolution of all nine components of our hypothetical sample, as shown in Figure 2.1c. What lie beyond this central concept of 2D separation are three major challenges

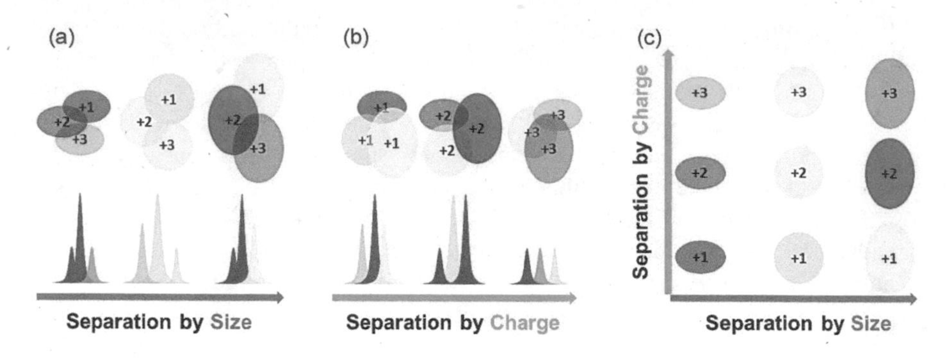


**FIGURE 2.1** Simple illustration of the attractiveness of 2D separations in general. For this hypothetical mixture of nine compounds, which vary in both size and charge, a conventional 1D separation by either size (a) or charge (b) yields a series of peaks that contain sub-mixtures of compounds for which the separation is not selective. However, when these complementary separations are combined in a 2D separation format (c), the mixture can be fully resolved in the same analysis time that was needed for the 1D separations.

that experimentalists must address to realize the benefit of 2D separation in practice: (1) finding two highly complementary separation mechanisms that are suitable for the sample at hand; (2) transferring $^1$D effluent to the $^2$D column in a way that both avoid re-mixing analytes already separated by the $^1$D column and avoid compromising the separation performance of the $^2$D column; and (3) optimizing the separation conditions to satisfy goals for the resolution of one or more groups of analytes and for detection sensitivity. In this overview section, we will briefly touch on each of these challenges. Readers interested in learning about them in more detail are encouraged to consult a number of recent discussions of the state of the art in these areas [11–14].

### 2.3.1 Modes of Two-Dimensional Separation

The central concept of 2D separation is that analytes not separated by a $^1$D column can be transferred as a mixture to a $^2$D column for further separation. However, the scope of this transferring process can vary greatly, ranging from a single transfer of $^1$D effluent containing analytes of particular interest to multiple, serial transfers across the entire $^1$D separation. We refer to the former mode as *heart-cutting* 2D separation (LC-LC) and the latter as *comprehensive* 2D separation (LC × LC) [15]. While LC-LC and LC × LC have been known since the late 1970s, more recently, two additional variants of 2D separation have been introduced and explored, namely *multiple heart-cutting* 2D separation (mLC-LC) and *selective comprehensive* 2D separation (sLC × LC). A brief discussion of the ways in which the four modes of 2D separation are related to each other is warranted here.

Figure 2.2 illustrates the basic concept of each mode of operation. Figure 2.2a shows that simple LC-LC involves the transfer of a single fraction of $^1$D effluent to the $^2$D column for further separation. The advantage of this approach is that the implementation is quite straightforward in practice. The resulting 2D separations can be very effective for the resolution of one or a few target analytes from complex matrices. Such separations have been demonstrated since the 1970s in application

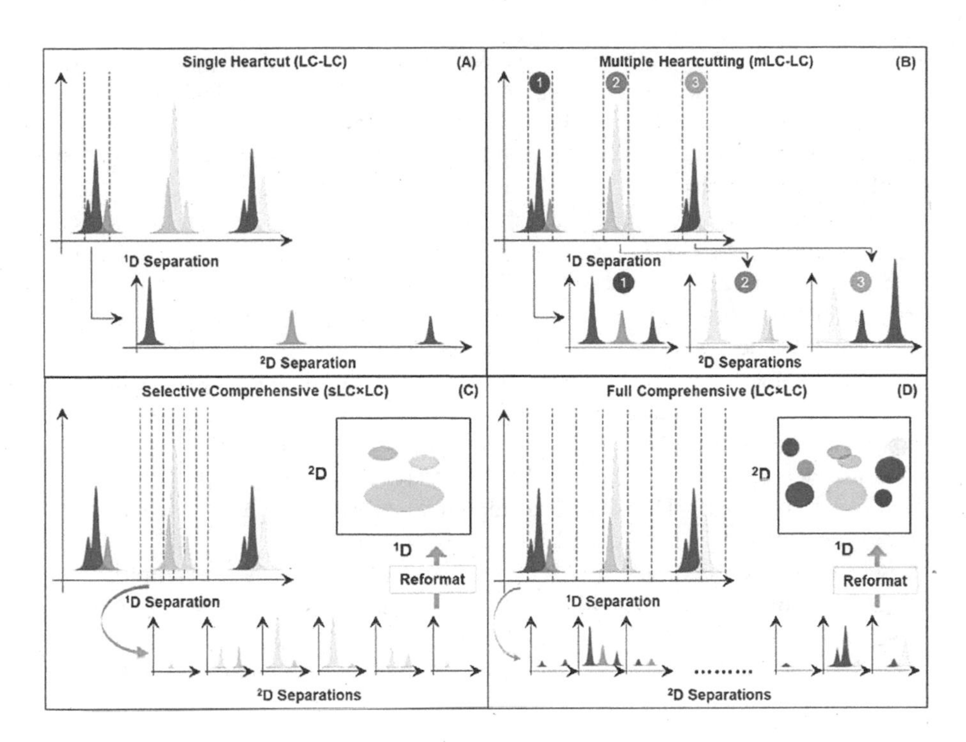


**FIGURE 2.2** Illustration of four different modes of 2D-LC separation. (a) Single heart-cut (LC-LC); (b) Multiple heart-cutting (mLC-LC); (c) Selective Comprehensive (sLC × LC); (d) Full Comprehensive (LC × LC).

areas ranging from PK studies to environmental analyses [13]. This approach has also been extended to three dimensions of separation [16].

The principle of mLC-LC—illustrated in Figure 2.2b—is to extend the single heart-cut approach, so that multiple regions of the $^1$D separation can be targeted, with single transfers of $^1$D effluent in each targeted region [17,18]. This approach is particularly powerful when one uses a sampling interface that allows sampling of the $^1$D separation and subsequent $^2$D separation of the transferred fractions to be executed in parallel. In principle, this allows the user to target as many regions of the $^1$D separation as needed, provided that there is sufficient spacing between regions of interest.

Moving on to the sLC × LC mode of 2D separation, Figure 2.2c shows that the idea that differentiates mLC-LC and sLC × LC is that in sLC × LC, multiple fractions of $^1$D effluent are transferred to the $^2$D column in a particular region of interest. There are two main motivations for executing the 2D separation in this way. First, transferring multiple small fractions of $^1$D effluent in a region of interest preserves any resolution of target analytes developed by the $^1$D separation [19,20]. In contrast, in the simple single heart-cut case, taking a single large fraction of $^1$D effluent can mix adjacent peaks that had already been separated by the $^1$D column back together in the transfer process. Obviously, this can be counterproductive. The second motivating factor is that taking multiple small fractions gives the user more flexibility with respect to the interplay between the sample loop volume used in the interface and the $^1$D flow rate. For example, if we have a $^1$D flow rate of 500 μL/min and we wish to sample a region that is 30-second wide, doing so in a single heart-cut mode requires a sample loop that is on the order of 300 μL. On the other hand, if we have an interface that allows us to capture and transfer five fractions in this region, then a set of five 60-μL loops would suffice. In general, using smaller loops enables faster $^2$D separations and reduces the effects of sample volume overload on the performance of $^2$D column.

Finally, Figure 2.2d illustrates the principle of LC × LC separation, which is the extreme opposite of the single heart-cut implementation. The principle here is that all of the $^1$D effluent is transferred to the $^2$D column at some point or at least a consistent fraction of it (e.g., when the flow is split after the $^1$D column but prior to injection into the $^2$D column), through repeated serial transfers over the entire course of the $^1$D separation. The primary advantage of this approach is that it provides the user with a holistic view of what is in the sample, from the point of view of elution from both the $^1$D and $^2$D columns. This additional analytical information naturally comes at the cost of increased instrument complexity and operational cost.

As will be discussed below in Sections 2.4–2.8, there are now examples in the literature involving 2D separations of mAb and related materials using each of the four modes described previously. At the start of method development, it is most important for users to consider which of the four 2D separation modes is best suited to the application at hand.

### 2.3.2 Basic Mechanics of Coupling Two Separations Together in a Two-Dimensional Format

A representative diagram showing the components common to most instruments used for 2D-LC is shown in Figure 2.3. Many variations of this setup can be found in the literature, as is evident from Table 2.2. The most comprehensive review of the different valve configurations that have been used was assembled by Mondello [21]. The first dimension of most 2D-LC systems looks very similar to a conventional 1D-LC instrument. To this is added some kind of interface to connect the first and second dimensions of the system together. In Figure 2.3, this interface is an 8-port/2-position valve that is fitted with two nominally identical sample loops, L1 and L2. Effluent leaving the $^1$D column enters one of these two loops (in the left image of the valve, loop L1) for some predetermined length of time. At the end of this time (i.e., the sampling time), the valve is switched (see red arrow in Figure 2.3 that indicates the flow paths for the two positions of the valve), such that the $^1$D effluent that was collected in loop L1 enters the $^2$D flow stream and effectively injected into the $^2$D column. The interface valve shown here is commonly used for LC × LC separations; this interface can also be used for simple LC-LC implementations. The mLC-LC and sLC × LC modes of 2D separation require a more sophisticated interface, where the loops L1 and L2 are replaced with multi-loop sampling valves that enable separation of the processes of sampling the $^1$D separation and injection of the fractions into the $^2$D column. Such interfaces have been described in detail elsewhere [17–19,22,23]. Within the past several years, there has been a great deal of research

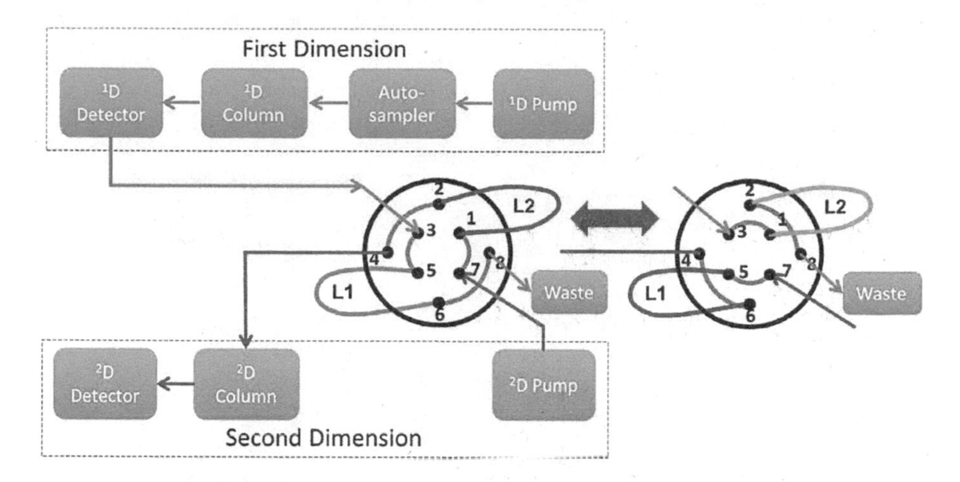


**FIGURE 2.3** Diagram of a typical 2D-LC instrument configuration highlighting the role of the valve between the $^{1}$D detector and the $^{2}$D column that captures fractions of $^{1}$D effluent and injects them into the $^{2}$D column. The red arrow indicates that the change in valve position enables switching the role of each sampling loop between the collection of $^{1}$D effluent and injection into the $^{2}$D column. (Reprinted from Stoll, D.R. and Carr, P.W., *Anal. Chem.*, 89, 519–531, 2017. With permission.)

focused on improving the performance of the interface by replacing or augmenting the valve-based type of interface shown here; these technologies have been discussed in detail recently by Egeness and coworkers [24], and Stoll [13].

### 2.3.3 Selecting Separation Mechanisms

As illustrated earlier in this section, a critical step in the development of a useful 2D separation involves choosing which separation mechanisms will be used in the first and second dimensions of the 2D system. The most thorough treatment of this topic to date appeared recently from Pirok and Schoenmakers [12], and readers interested in a more detailed discussion of this topic will find their review article very useful. In Table 2.1, we have adapted their "red light/green light" approach to visualizing which combinations of separation mechanisms are likely to be useful for separations of mAbs and related materials. Methods involving combinations highlighted green (e.g., RP × RP) are likely to be more straightforward to develop and highly effective, whereas combinations highlighted in red are likely to be difficult to develop, and ineffective. Those combinations with intermediate colors may or may not be effective, depending on the sample at hand and the goals of the method. As with all summaries of this kind, this table is intended to provide guidance and should not be viewed as definitive. As technology for 2D-LC continues to develop, we expect that this table will evolve toward more shades of green from shades of red.

**TABLE 2.1**
**Performance Potential of Different Pairings of Separation Mechanisms for 2D Separations of mAbs and Related Materials**

| | | $^2$D | | | | | |
|---|---|---|---|---|---|---|---|
| | | RP | HILIC | HIC | IEX | SEC | Affinity |
| $^1$D | RP | | | | | | |
| | HILIC | | | | | | |
| | HIC | | | | | | |
| | IEX | | | | | | |
| | SEC | | | | | | |
| | Affinity | | | | | | |

*Source:* Pirok, B.W.J. et al., *J. Sep. Sci.*, 2017.

Discussion of a few examples is warranted here to illustrate the kind of thinking that lies behind the table. First, consider the combination of reversed-phase (RP) separations in both dimensions. The potential advantages of this combination are numerous, including high column efficiencies, availability of many commercial columns, and compatibility with MS. On the surface, it appears that these two separations would not be complementary in the same sense as that illustrated by Figure 2.1. However, eluent pH can be used to effectively induce complementarity of two RP separations by changing the ionization state of analytes and the way they interact with the stationary phase, especially for certain classes of molecules such as peptides [17,18]. As a result, this type of method has been very successful for the characterization of mAbs at the peptide level, as will be discussed in Section 2.4.3. Second, consider the combination of ion-exchange (IEX) and RP separations. On one hand, IEX separations are not known for high efficiency as RP separations are, and so, this combination may not appear as attractive as the RP × RP combination. However, the IEX × RP combination is particularly attractive for other reasons. First, the highly aqueous eluents that are typically used for IEX separations are very attractive, because large volumes of $^1$D IEX effluent can be injected into $^2$D RP columns without negatively affecting peak shapes and widths in the $^2$D separation. Second, IEX columns are exquisitely selective for proteins varying only by a single charge, adding a level of selectivity that simply cannot be accessed using an RP separation alone. In other words, IEX and RP are highly complementary from the point of view of separation selectivity. Finally, consider the combination of hydrophobic interaction chromatography (HIC) with IEX, which carries the yellow designation in Table 2.1. The main concern here is that the high-ionic-strength mobile phases (e.g., >1 M salt) required for HIC separations of proteins may have negative effects on the $^2$D IEX separation. Specifically, the high salt content of the $^1$D HIC effluent, which becomes the sample matrix for fractions injected into the $^2$D IEX column, will result in no retention at all and thus no $^2$D resolution unless something is done to counteract this effect. The yellow designation here does not indicate that coupling HIC and IEX is impossible; rather, it indicates that method development has to be approached carefully and thoughtfully to produce an effective 2D method. The final point we will make here about choosing separation mechanisms is that the ordering of the mechanisms can, and often does, matter. In some cases, the preferred ordering is more obvious than in other cases. For example, when MS detection is used following the 2D separation, it is obviously beneficial to use the more MS-friendly separation mechanism in the second dimension. For example, when coupling IEX and RP, the best option is to use the IEX separation in the first dimension and the RP separation in the second dimension. In other situations, $^{2}$D separation speed may be a factor, and re-equilibration of the $^{2}$D column after each $^{2}$D separation cycle can be particularly important. While it has been shown that re-equilibration of RP columns used for the separation of small molecules with solvent gradient elution conditions can be extremely fast (i.e., about one column volume required to achieve repeatable retention times) [25–27], it seems likely that other columns may not re-equilibrate as quickly.

### 2.3.4 Avoiding Loss of $^{1}$D and $^{2}$D Separations

The benefits of 2D separation over conventional 1D separation, as illustrated in Figure 2.1, are straightforward to understand. Realizing these benefits in practice, however, requires attention to several important experimental details. Foremost among these are method development decisions that impact the resolution that the $^{1}$D and $^{2}$D separations contribute to the overall 2D resolution. Specifically, we must be mindful that coupling the two individual separations does not significantly diminish the inherent performance of each separation.

Resolution provided by the $^{1}$D separation can be diminished by a variety of sources of extra-column peak dispersion such as injection volume broadening and peak dispersion in connecting tubing. All of the theory we have from conventional 1D separations applies just as well to the $^{1}$D separation of a 2D system as it does to a simple 1D separation. However, there is an additional impact on the $^{1}$D resolution that arises in the context of 2D separations—this is the so-called effect of undersampling [28,29]. Figure 2.3 illustrates the effect of the rate at which the $^{1}$D separation is sampled on the effective resolution that the first dimension contributes to the 2D separation. These chromatograms are the results of simulations that show the reconstruction of the $^{1}$D chromatogram after peaks have been broadened by undersampling [13]. Panel A shows that when fractions of the $^{1}$D effluent are transferred to the $^{2}$D column such that the volume of each fraction ($V_s$) is small compared with the $^{1}$D peak volume ($^{1}V_p$), analytes already separated by the $^{1}$D column remain separated as they proceed to be injected into the $^{2}$D column. As the fraction volume is increased relative to the peak volume, however, re-mixing of already-separated analytes begins to occur, and this effectively diminishes the resolution provided by the $^{1}$D separation. Panel C quantifies this effect by relating the percentage increase in $^{1}$D peak width to the ratio of the fraction volume to the peak volume. We see that even when the sampling volume is one-half of the peak volume, the width of $^{1}$D peaks doubles just owing to undersampling.

Since the peak and fraction volumes are related through the $^1$D flow rate, we can also discuss this problem in terms of the ratio of the sampling time (i.e., the time over which $^1$D effluent is collected prior to injecting the fraction into the $^2$D column) to the $^1$D peak width in time units. This is convenient because the sampling time is related to the time devoted to each $^2$D separation cycle (i.e., the cycle time). Indeed, in the case of LC × LC, these two times must be equal. With only Figure 2.4 in mind, one would be tempted to conclude that we should use very short sampling times, so as to not make the $^1$D peaks broader by undersampling them. The problem with this of course is that as we make the $^2$D cycle time short, we sacrifice the degree of resolution that the $^2$D separation can contribute to the 2D separation. These opposing effects naturally lead to an optimum value of the $^2$D cycle time that minimizes the effect of undersampling while maximizing the resolution of each $^2$D cycle to the extent possible within these constraints [30]. This idea of an optimal $^2$D cycle time has been studied extensively for 2D separations of small molecules and found to be about 15 seconds [31–33], whereas for peptides and proteins, this is an active area of research [34].

Turning to the effect of coupling two separations together on the second dimension, here too, there are method development decisions that can have a big impact on the resolution obtained from each $^2$D separation cycle. Figure 2.5 shows the potential seriousness of this issue, using conditions that are relevant to the second dimension of LC × LC separations in particular. Panel A shows the chromatogram obtained for a mixture of five alkylphenone homologs injected into a 30-mm × 2.1-mm Inner Diameter (i.d.) C18 column and eluted with a solvent gradient running from 50% to 90% Acetonitrile (ACN)

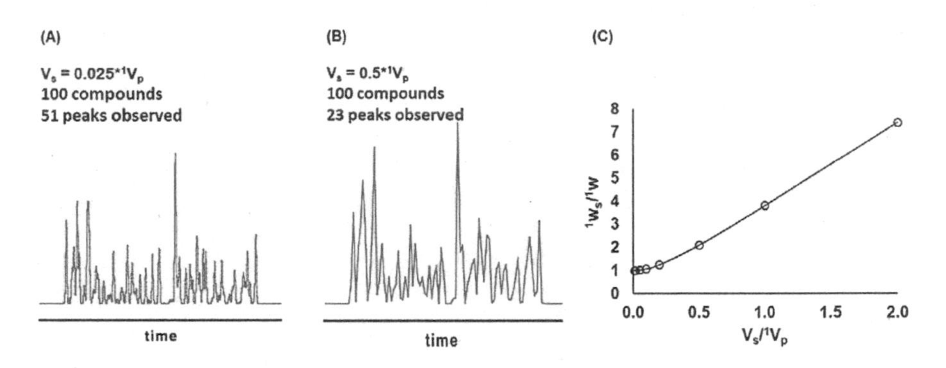


**FIGURE 2.4** Effect of undersampling on widths of $^1$D peaks. Panel (A) shows a simulated $^1$D separation of a mixture of 100 compounds, with a sampling volume that is 1/40th of the $^1$D peak width prior to sampling. In this case, 51 peaks are observed in the $^1$D chromatogram. Panel (B) shows how seriously the $^1$D resolution is degraded when we simulate the same separation but with a sampling volume that is more representative of what we can achieve in practice in real 2D separations; the number of observed peaks decreases by over 50%. Panel (C) shows the relationship between the widths of $^1$D peaks before ($^1$w) and after ($^1w_s$) sampling and the ratio of the sampling ($V_s$) and $^1$D peak volumes ($^1V_p$), where the peak volume is defined as the 8*σ peak width in volume units. The trend in (C) is calculated from eq. 10 of ref. [29], with $\kappa = 0.21$.

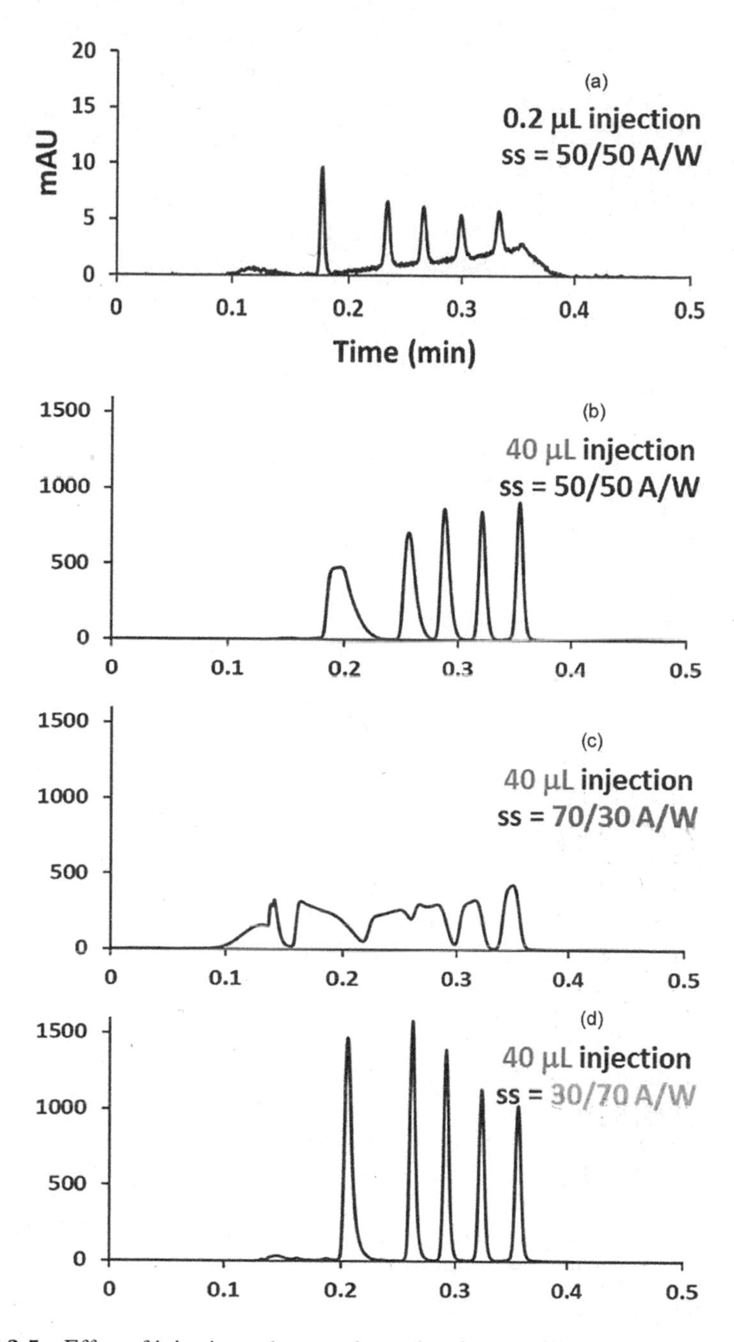


**FIGURE 2.5** Effect of injection volume and sample solvent on 2D separation performance. Injection volumes and sample solvents are (a) 0.2 µL and 50/50 ACN/water; (b) 40 µL and 50/50 ACN/water; (c) 40 µL and 70/30 ACN/water; and (d) 40 µL and 70/30 ACN/water. Other conditions: column, 30 mm × 2.1 mm i.d. Zorbax SB-C18 (3.5 µm); flow rate, 2.5 mL/min; temperature, 40°C; and gradient elution from 50% to 90% ACN in 15 seconds, with water as the A solvent. Analytes are alkylphenone homologs, all at 1 µg/mL.

over 15 seconds. The injection volume in this case was 0.2 μL, and the analytes were dissolved in 50/50 ACN/water. All the peaks have very nice shapes, but the injection volume is too small relative to the $^2$D column volume to be very useful in the context of 2D-LC. Increasing the injection volume to 40 μL, but with the same sample, results in the chromatogram in Panel B. Here, the detection sensitivity is much better, simply because more analyte mass is injected into the column, but now, the resolution is compromised, particularly for the early peaks, because of the influence of the high level of ACN in the sample. If we now consider the worst-case scenario (Panel C), where the sample contains more ACN than the starting eluent in the gradient elution program—a situation that is not uncommon in 2D-LC—resolution of the sample is almost entirely lost, again owing to the strong influence of the sample solvent on the separation. Now, if we have the ability to influence the sample solvent composition during the transfer of the $^1$D effluent fraction to the $^2$D column, we can dramatically improve the situation. Panel D shows that reducing the ACN content by roughly 50% results in a chromatogram with both high detection sensitivity and good resolution of the sample. The importance of this issue has motivated a number of groups to study potential solutions to the problem [13,24]. Indeed, this was the driving force behind our development of the Active Solvent Modulation (ASM) concept that we described recently [22]. This is another aspect to consider when thinking about coupling different separation mechanisms, as described in Section 2.3.3.

### 2.3.5 Optimizing for Resolution, Speed, and Sensitivity

The topics discussed so far in this section have been mainly conceptual in nature in that understanding the ideas provides a foundation for the basic structure of a 2D method (i.e., choosing a mode of 2D separation and choosing separation mechanisms). In most cases, though, we have separation goals that are at least semi-quantitative, and working toward these goals requires consideration of more details as part of the optimization of the method. For example, in targeted applications focused on one or a few analytes, we are typically interested in resolving the targets from interfering compounds and reaching some predetermined level of detection sensitivity (i.e., detection limits). In other applications that are more open-ended, maximizing peak capacity or resolving power of the method within a particular analysis time may be the primary goal. In all of these cases, there are many decisions to be made, including physical dimensions of the $^1$D and $^2$D columns, particle sizes, particle morphologies, flow rates, fraction volume, sampling time, and so on. A thorough discussion of these details is beyond the scope of this chapter. Fortunately, there are now several excellent resources to support the optimization of LC × LC methods, with an eye toward managing the trade-off between peak capacity and detection sensitivity [12,13,34–37]. For the newer mLC-LC and sLC × LC modes of 2D separation, systematic optimization schemes have not been developed yet, but these are likely to emerge soon, as the body of work involving these methods continues to grow.

## 2.4 CHARACTERIZATION OF MONOCLONAL ANTIBODIES

### 2.4.1 Overview

Characterization of mAbs presents both significant challenges and opportunities to state-of-the-art analytical methods. On the one hand, the molecules are very large (ca. 150 kDa) and are often presented to the analyst as highly heterogeneous mixtures, both in the form of "pure" protein and in the form of materials obtained at different stages in the protein expression (e.g., N- and O-glycosylation) or process development. In the case of mAb drug substance, structural "hot spots" may be modified via multiple degradation pathways. In the case of unpurified material, samples containing not only the mAb of interest but also impurities such as other host cell expression products (e.g., DNA), HCPs, and host cell culture media components are presented to the analyst. These sample complexities do not lend themselves to definitive characterization using a single technique. Rather, a large suite of complementary and orthogonal techniques is required to assemble as complete a picture of the mAb characteristics as possible.

On the other hand, the structural complexity of mAbs provides opportunities to study the molecules at different levels of detail and with different foci. For example, glycosylation of the mAb can be studied at the intact protein level by determining the mass of the protein with all glycans attached. We can also study glycosylation at lower levels either by enzymatically releasing the glycans from the protein prior to the analysis, enabling analysis of both the free glycans and the glycan-free protein, or by enzymatically cleaving the amino acid backbone of the protein to produce a series of glycopeptides that allows the determination of specific glycosylation sites.

In this section, we have organized our discussion of recent work on the use of 2D-LC for the analysis of mAbs by the level of analysis—intact/subunit/fragment-level and peptide-level analysis (Table 2.2).

### 2.4.2 Intact and Subunit/Fragment Levels

As stated previously, a consequence of the complex nature of mAbs and related materials is that no single analytical technique is adequate to fully characterize these materials. In many cases, an analytical approach will involve the interrogation of the sample at different levels of the structure and/or chemistry [8]. Figure 2.6 shows examples (i.e., many other enzymes and reducing agents can be used) of different sample pretreatment steps that can be used to chemically disassemble the mAb prior to analysis. These approaches produce mixtures of products that enable analysis at different levels, ranging from the intact, native structure (i.e., glycosylated, with non-covalent associations present) down to the peptide level.

#### 2.4.2.1 Two-Dimensional Separation by Affinity and Size

One attractive implementation of 2D-LC at the intact protein level involves the use of an affinity-based separation (e.g., protein A) in the first dimension, followed by a

**TABLE 2.2**
**Summary of All Literature on 2D-LC Separations of mAbs and Related Materials**

| First Author | Year | References | Separation Focus | Offline/ Online | 2D Separation Mode | Interface | $^{1}$D Separation | $^{2}$D Separation |
|---|---|---|---|---|---|---|---|---|
| Ehkirch | 2018 | [38] | Identification of ADC species under native conditions | Online | LC × LC | Dual 6-port valves | HIC | SEC |
| Gstöttner | 2018 | [39] | Identification of mAb microvariants | Online | mLC-LC | Multiple heart-cutting | CEX | RP |
| Gilroy | 2017 | [40] | ADCs | Online | LC-LC | 6-Port/2-position valve | HIC | SEC or RP |
| Williams | 2017 | [41] | mAb titer and aggregation | Online | sLC × LC | Dual 12-position selection valves | Protein A | SEC |
| Sandra | 2017 | [42] | Versatility of 2D-LC methods for mAbs | Online | LC-LC and LC × LC | 8-Port/2-position valve | Protein A, RP | SEC, CEX, RP |
| Yan | 2017 | [43] | Aggregation level of mAb charge variants | Online | mLC-LC | Multiple heart-cutting | CEX | SEC |
| Luo | 2017 | [44] | Identification of impurity in therapeutic peptide | Online | LC-LC | 8-Port/2-position valve | RP | RP |
| Largy | 2016 | [45] | Versatility of 2D-LC methods for mAbs and ADCs | Online | LC-LC | | CEX, SEC, HIC | RP |
| Sorensen | 2016 | [46] | Comparison of originator/ biosimilar mAbs | Online | LC × LC | 8-Port/2-position valve | CEX | RP |
| Zhang | 2016 | [47] | Detection of HCPs in study of downstream processing | Online | LC × LC | Dual 6-port valves | RP | RP |
| Sandra | 2016 | [48] | Study of drug conjugation in lysine-linked ADC | Online | mLC-LC, LC × LC | Multiple heart-cutting | CEX, RP | RP |
| Sarrut | 2016 | [49] | Optimization of HIC × RP separation of ADCs | Online | LC × LC | 8-Port/2-position valve | HIC | RP |

*(Continued)*

**TABLE 2.2 (*Continued*)**
**Summary of All Literature on 2D-LC Separations of mAbs and Related Materials**

| First Author | Year | References | Separation Focus | Offline/ Online | 2D Separation Mode | Interface | $^1$D Separation | $^2$D Separation |
|---|---|---|---|---|---|---|---|---|
| Sarrut | 2016 | [50] | Identification of drug-linked mAb subunits in ADCs | Online | LC × LC | 8-Port/2-position valve | HIC | RP |
| Li | 2016 | [51] | Detection of free drug in ADC | Online | LC-LC | Dual 6-position selection valves | SEC | RP |
| Heudi | 2016 | [52] | Detection of free drug from ADC in serum | Online | LC-LC | 6-port/2-position valve | RP | RP |
| Farrell | 2015 | [53] | Detection of HCPs | Offline | LC × LC | N/A | RP | RP |
| Birdsall | 2015 | [54] | ADCs | Online | LC-LC | Dual 6-port valves | HIC | RP |
| Doneanu | 2015 | [55] | Detection of HCPs | Online | LC × LC | Dual 6-port valves | RP | RP |
| Shen | 2015 | [56] | Peptide-level detection of mAb in serum | Online | LC-LC | 6-Port/2-position valve | RP | RP |
| Vanhoenacker | 2015 | [57] | Peptides | Online | LC × LC | 8-Port/2-position valve | CEX, RP, HILIC | RP |
| Li | 2015 | [58] | Detection of free drug and impurities in ADC | Online | LC-LC | Dual 6-port valves | SEC | RP |
| Wang | 2015 | [59] | Determination of degradant in mAb formulation | Online | LC-LC | 8-Port/2-position valve | SEC | HILIC |
| Stoll | 2015 | [23] | Direct identification of mAb charge variants | Online | sLC × LC | Multiple heart-cutting | CEX | RP |
| Li | 2014 | [60] | Degradation of polysorbate in mAb formulation | Online | LC-LC | Dual 6-port valves | Mixed-mode AEX/RP | RP |
| St. Amand | 2014 | [61] | At-line monitoring of mAb charge variants during cell culture | Online | LC-LC | Dual 6-port valves | Protein A | CEX |

(*Continued*)

**TABLE 2.2 (*Continued*)**
**Summary of All Literature on 2D-LC Separations of mAbs and Related Materials**

| First Author | Year | References | Separation Focus | Offline/ Online | 2D Separation Mode | Interface | $^1$D Separation | $^2$D Separation |
|---|---|---|---|---|---|---|---|---|
| Zhang | 2014 | [62] | Tracking of HCPs during downstream processing | Online | LC × LC | Dual 6-port valves | RP | RP |
| Doneanu | 2012 | [63] | Detection of HCPs | Online | LC × LC and LC-LC | Dual 6-port valves | RP | RP |
| He | 2012 | [64] | Quantitation of protein and excipients in biopharmaceutical products | Online | LC-LC | 6-Port/2-position valve | SEC | Mixed-mode AEX/CEX/RP |
| Mazur | 2012 | [65] | Identification of mAb degradation products | Offline | LC × LC | N/A | SEC | RP |
| Alvarez | 2011 | [66] | Identification of charge variants | Online | mLC-LC | Dual 6-position selection valves | CEX, SEC | RP |

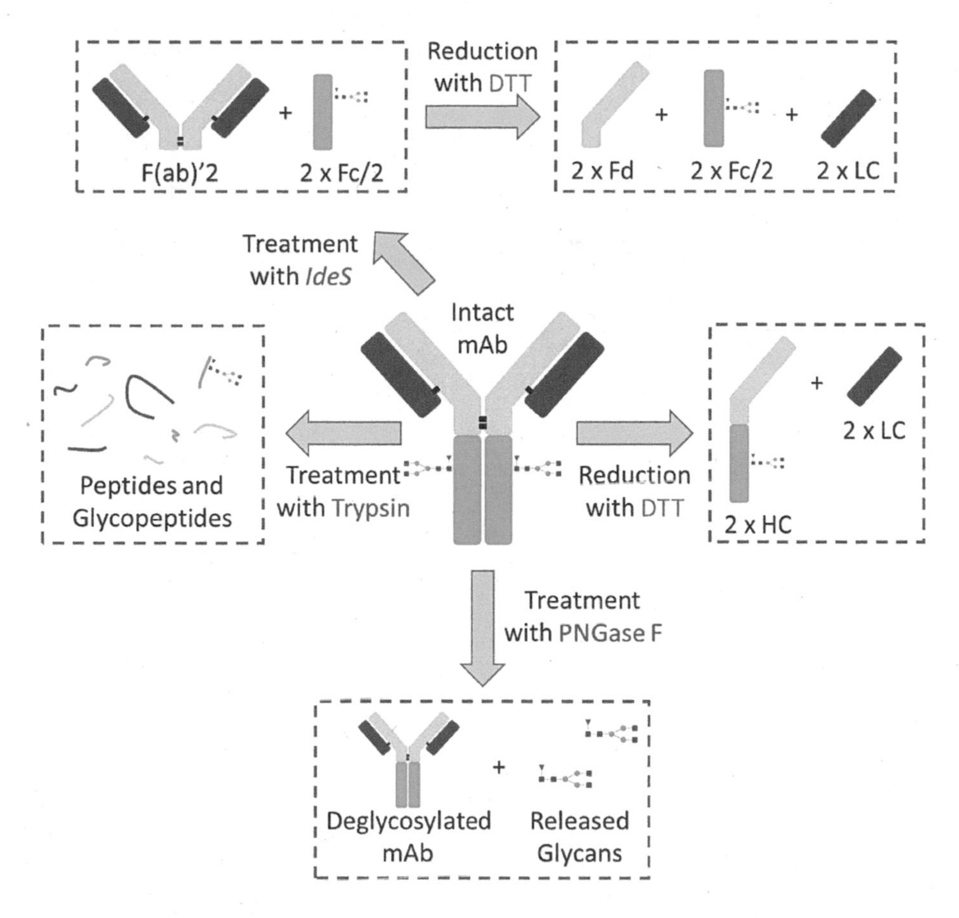


**FIGURE 2.6** Summary of different approaches to sample pretreatment of mAbs prior to analysis by liquid chromatography.

sized-based separation (e.g., SEC) in the second dimension [41,42]; Figure 2.7 shows results from the recent work of Sandra and coworkers along these lines. In this implementation, the $^{1}$D separation can be used to retain and separate mAb from complex materials, such as host cell supernatant, with extraordinary selectivity (Figure 2.7C). At this point, the concentration (titer) of mAb can be easily determined by using ultraviolet (UV) absorbance detection by comparison to a calibration curve constructed, using mAb standards of known concentration (e.g., see Figure 2.7B). Then, one [42] or multiple [41] fractions of the $^{1}$D mAb peak are transferred to a $^{2}$D SEC separation, where mAb monomer is separated from both low (e.g., clipped antibody structures and degraded protein) and high (e.g., dimers and higher-level aggregates) molecular weight species present in the sample (e.g., see Figure 2.7F). The demonstration of this method from Sandra and coworkers showed how this 2D-LC approach can be used to rapidly screen several candidate host cell clones for ones that produce both a high supernatant concentration of mAb and a low level of mAb aggregation.

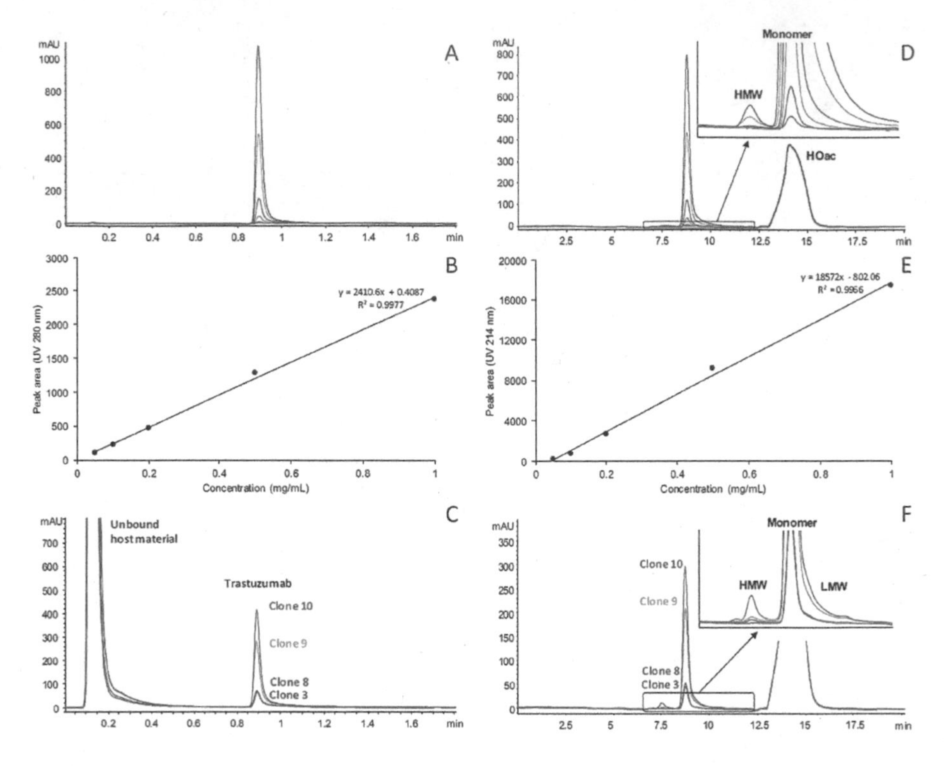


**FIGURE 2.7** Demonstration of the use of LC-LC with protein A affinity-based separation in the first dimension for the determination of mAb concentration (A/C), followed by SEC separation in the second dimension for the determination of the level of mAb aggregation (D/F). Panels B and D are calibration curves indicating the relationship between UV detection peak area and mAb concentration following the $^1$D and $^2$D separations, respectively. (Reprinted from Sandra, K. et al., *J. Chromatogr. A*, 1523, 283–292, 2017. With permission.)

### 2.4.2.2 Two-Dimensional Separation by Charge and Size

While the combination of protein A and SEC separations described in the previous section provides an efficient and effective means of determining mAb concentration and the overall level of aggregation, it cannot provide information about the distribution of aggregation over different charge variants of the mAb. For this purpose, Yan, Chen, and coworkers developed an mLC-LC separation involving cation-exchange (CEX) separation in the first dimension, followed by SEC separation in the second dimension. An MS-incompatible buffer was used for the SEC separation, and thus, no identifying information was obtained for the mAbs from this analysis. However, the results of the 2D separations did show that the majority of mAb aggregates in the sample were observed eluting later in the CEX separation compared with the primary form of the mAb, meaning that most of the aggregation occurred in basic (i.e., higher pIs) variants of the protein. This is another good example of the ways in which 2D-LC can accelerate understanding these complex molecules by obtaining orthogonal analytical information in a single time-efficient analysis.

### 2.4.2.3 Two-Dimensional Separation by Charge and Subunit

CEX separations have been used extensively to assess the charge variant profile of mAbs [9,67–69]. One challenge with this type of work, however, is that the mobile phases typically used for this work often contain too much non-volatile salt to be compatible with direct detection using MS. Alvarez, Borisov, and coworkers demonstrated an early application of mLC-LC 2D separation involving a $^1$D separation by CEX, followed by a $^2$D RP separation and subsequent MS detection [66]. In our own work, we recently expanded the scope of this type of separation to the LC × LC separation format and demonstrated the utility of this type of separation through the comparison of three different pairs of originator and biosimilar mAbs varying in their degree of similarity [46]. Since the $^2$D RP separation is compatible with MS detection, we obtain from these analyses the charge variant profile and masses of each of the separated subunits, all in a 1-hour analysis. As a representative example from this work, Figure 2.8 shows that some of the detected subunits (i.e., Fc/2 and F(ab')$_2$ in Figure 2.6) are indistinguishable (e.g., the Fc/2 subunit, panels E/F),

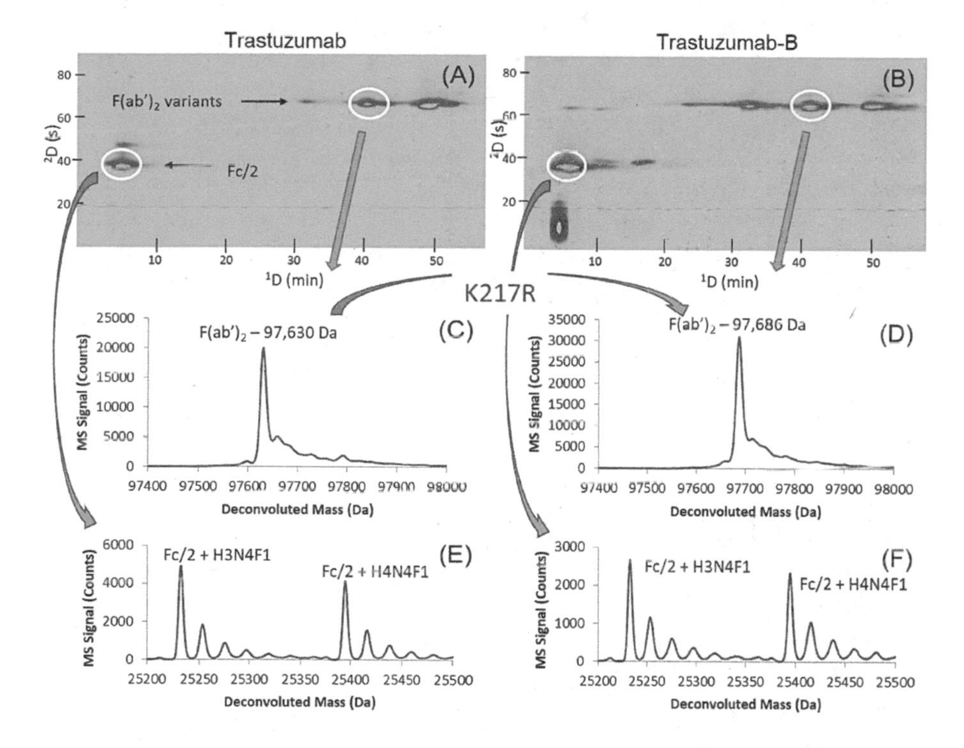


**FIGURE 2.8** LC × LC-MS separations of *IdeS* digested trastuzumab (A) and biosimilar candidate trastuzumab-B (B), with cation-exchange and reversed-phase separations in the first and second dimensions, respectively. Panels C and D show that a one-amino-acid difference in the F(ab')$_2$ subunits of these two proteins is easily detected, even though they elute in the same region of the 2D chromatograms. On the other hand, other parts of the molecules that are known to be very similar, such as the Fc/2 subunits, appear very similar by MS, as shown in panels E and F. (Reprinted from Stoll, D.R. and Maloney, T.D., *LCGC N. Am.*, 35, 680–687, 2017. With permission.)

indicating that this part of the molecule is the same in the originator and biosimilar molecules. On the other hand, other parts of the molecule are found to be different, such as the $F(ab')_2$ subunit shown in panels C and D, where a one-amino-acid substitution from lysine to arginine was detected. This type of method provides a way to rapidly screen for small differences between molecules that are believed to be quite similar.

### 2.4.3 Peptide Level

Characterization of therapeutic proteins such as mAbs at the peptide or "bottom-up" level is a well-established approach for confirming product identity. In these experiments, the drug products are treated with proteases (see "Treatment with Trypsin" in Figure 2.6) to generate peptides, which are then separated and detected by using UV alone or in combination with MS. The principal benefit of peptide mapping is the ability to develop a deep understanding of site-specific levels of relevant modifications, such as deamidation, oxidation, and sequence variants. On the other hand, confounding artifacts resulting from lengthy sample preparation steps as well as per-sample analysis times that are typically long are also a factor. Currently, the combination of peptide mapping experiments with MS is gaining popularity for process development and as a stability indicating assay in the form of the emerging multi-attribute method (MAM) [70,71].

In their recent work, Sandra et al. applied LC × LC separations with RP columns in both dimensions to tryptic digests of originator tocilizumab and compared the peptide map with that obtained for the same molecule from two different Chinese hamster ovary (CHO) host cell clones [42]. As is shown in Figure 2.9 from their work, the high peak capacity achieved in the LC × LC separation readily revealed a unique peptide feature in one of the clones. Since the separation was performed on-line with tandem MS, it was possible to identify a Phe to Leu/Ile mutation, which accounted for a 68-Da mass shift (2 × 34 Da, accounting for each of the heavy chains) on the intact mAb. Aside from this difference, the approach verified other features among the three mAbs, such as the degree of C-terminal lysine processing, which correlated with IEX-based separations also performed in this study. The bottom-up 2D-LC approach used in this work provides a high-resolution snapshot of mAb structure, which can inform clone selection or process development.

Peptide-level 2D-LC was also applied to PK studies of therapeutic proteins, for example, the quantitation of immunoglobulin A1 protease (IgAP) in human serum [56]. A 2D-LC separation using RP columns in both dimensions—one at high pH and the other at low pH (referred to hereafter as high pH/low pH RP-RP)—coupled with multiple reaction monitoring (MRM) MS detection was shown to be superior to enzyme-linked immunosorbent assay (ELISA)-based methods, owing to the lack of interference from anti-drug antibodies present in the serum. In this work, 2D-LC was applied with the intent to detect a single, specific surrogate peptide, which would represent the concentration of the IgAP molecule. Such peptides might be present in PK studies in the ng/mL levels among a complex matrix of peptides from

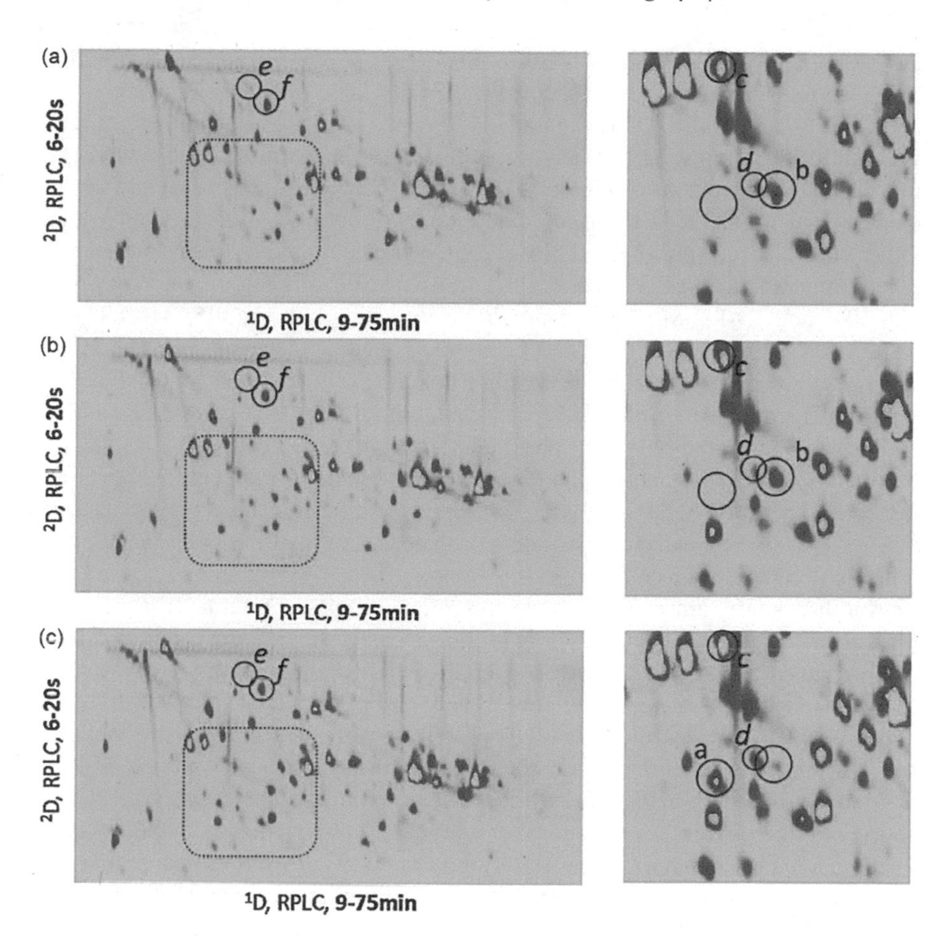


**FIGURE 2.9** High pH/low pH RPRP peptide maps of tocilizumab originator (a), tocilizumab from CHO clone A (b), and tocilizumab from CHO clone C (c). The peptide identified by **spot b** is NQFSLR, and the peptide identified by **spot a** is NQI/LSLR. (Reprinted from Sandra, K. et al., *J. Chromatogr. A*, 1523, 283–292, 2017. With permission.)

other serum proteins. The advantage that 2D-LC provided in this case was a 40-fold enhancement in sensitivity compared with 1D approaches. The authors went on to validate the assay for possible use for clinically relevant samples.

Luo et al. demonstrated the use of LC-LC for the characterization of impurities in therapeutic peptides [44]. The authors developed a $^{1}$D RP separation by using sodium perchlorate as an ion pairing reagent, which provided resolution of various impurity peaks. A $^{2}$D RP separation provided desalting of the MS-incompatible perchlorate and delivery of the impurity peak to the MS for mass measurement. A dehydrated variant of the therapeutic peptide was determined with high confidence owing to the $^{1}$D separation, which would not have been possible by using MS-friendly mobile phases. Positional isomers of a synthetic peptide were also characterized using the same strategy.

Two-dimensional LC separations have been used extensively in the area of proteomics for decades [72]. Nevertheless, as chromatography and instrument technologies evolve, the optimization of conditions for these types of separations must be revisited. The Heinisch group recently described the results of thorough optimization studies aimed at on-line LC × LC separations of peptides on the scale of several hours [37] and most recently for sub-hour separations [34]. From these studies, it is clear that LC × LC separations can increase the achievable peak capacity several fold over conventional 1D-LC separations, up to around 5,000 in an analysis time of 2 hours. Other groups have explored the limits of peak capacity for multi-dimensional separations at much longer analysis times. In a study by Spicer, Krokhin, and coworkers, 1D-, 2D-, and 3D-LC-MS separations with analysis times of 1.5, 31.5, and 189 hours, respectively, were compared for deep analyses of peptide mixtures [73]. The 2D-LC separation was based again on the use of RP columns in both dimensions and the high pH/low pH strategy. In the 3D-LC approach, a third dimension of RP separation was added, involving a low-pH mobile phase containing heptafluorobutyric acid as an ion-pairing agent. This third dimension was added prior to the first dimension of the existing 2D-LC system. The 3D approach enabled the identification of over 14,000 proteins in a single analysis, clearly demonstrating the power of the multiple orthogonal separations for deep study of a complex peptide sample.

Another study aimed to identify insecticidal toxin proteins from *Bacillus thuringiensis*, a bacterium commonly used as a pesticide [74]. Researchers used an on-line nanoscale mLC-LC system with CEX and RP separations in the first and second dimensions, respectively. The 2D separation was coupled to MS detection, and it enabled the identification of 15 protein toxins in this organism, including seven novel pore-forming crystal proteins. In a different study, Wang et al. developed an off-line high pH/low pH RP × RP separation and applied it to the identification of proteins from six common cell lines [75]. The $^{1}$D gradient, peptide sample loading, and on-line removal of interfering salts resulting from sample preparation steps were each carefully considered in order to maximize protein identification. Using this approach, it was possible to characterize the proteome of a given cell line (with between 7300 and 8500 protein identifications) within 1.5 days. The Borchers group developed an analytical scale, on-line version of the same type of method with MRM MS detection for plasma protein quantification [76]. This 2D approach greatly improved the sensitivity for a given plasma protein, with the majority of proteins enriched between 10- and 25-fold compared with 1D methods. Validation of the platform was successful, and the approach was applied to the automatic quantification of peptides representing 168 proteins of interest spanning seven orders of magnitude in concentration.

2D-LC has tremendous potential to improve workflows aimed at the characterization of therapeutic proteins at the peptide level. The sensitivity improvement over 1D-LC methods makes detection of low-abundance species straightforward, with applications to PK studies and proteomics. The additional peak capacity and selectivity provided by 2D-LC can be leveraged to rapidly identify differences between samples and support non-targeted workflows. Development of at-line assays to measure the attributes of protein therapeutics continues, and, here, the ability to quickly compare what may be minor differences in complex samples, is an area of growing interest. 2D-LC appears ideally suited for this task, and increasingly robust commercial systems supporting flexible implementations of a variety of 2D-LC methods are likely to provide a critical analytical capability for these process analytical technology (PAT) applications, ultimately enabling product attribute control (PAC) [70].

## 2.5 CHARACTERIZATION OF HOST CELL PROTEINS

HCPs represent a class of process-related impurities resulting from the expression of therapeutic proteins in cell culture. Even low levels of HCPs can affect the safety or efficacy of therapeutic protein products, so significant effort is made in designing manufacturing processes that efficiently clear HCPs from drug products [77]. By their nature, HCPs display a range of physicochemical properties. Adding to the challenge is that some HCPs are known to co-purify with the drug substance. Sensitive analytical assays are thus required to support the development of HCP clearance approaches in downstream processing, in addition to serving as a final product release assay. ELISA remains the gold standard method for HCP detection, but orthogonal methods such as 1D-LC-MS continue to gain traction [78].

When dealing with HCPs, analysts are faced with detecting low levels of impurities in the presence of a large excess of drug product (e.g., mAb). Typically, parts per million-level detection of HCPs is desired (nanogram of HCP per milligram of drug product). There are many parallels between HCP analysis and proteomics, but the former can generally be considered a simpler case, since the number of proteins in question is lower (generally $<100$ after protein A purification of an mAb). In either case, dynamic range is a major analytical challenge, making technologies that can drastically increase peak capacity, such as 2D-LC, appealing for the analysis of HCPs.

In recent years, the majority of 2D-LC-based HCP work has been aimed at improving the understanding of downstream processing steps. The influence of different protein A elution buffers on HCP clearance was demonstrated by Farrell, Bones, and coworkers [53]. In this work, digested mAb samples were fractionated using a $^{1}$D RP column operated at high pH. Fractions were then combined off-line by using a concatenation strategy [79] and injected on a $^{2}$D nanoscale RP column operated at low pH and coupled directly to a quadrupole-time-of-flight (QTOF) MS detector. A total of 40 unique HCPs were identified from an experiment involving four unique protein A elution buffers. For the particular mAb used in the study, the use of 100 mM arginine at pH 3.5 resulted in the fewest number of HCPs as well as the lowest overall quantity of HCPs compared with other investigated elution buffers. The impact of cell culture duration was also investigated in this study, and the authors demonstrated the detection of intracellular-derived HCPs, which correlated with declining cell viability.

Scientists at Amgen expanded on their earlier work [62], using 2D-LC-MS for HCP characterization. The first study applied this technique to spike challenge studies, where HCP clearance via a polishing step was studied using samples that had been enriched in HCP content by using anion-exchange (AEX) chromatography [80]. The authors demonstrated HCP clearance (to below 100 ppm) via chromatographic polishing steps when the HCP load was an order of magnitude above control levels. In addition, the redundancy of serial chromatographic polishing steps could be assessed, providing an understanding of the contribution of each step to the overall risk. The authors concluded that 2D-LC-MS was an essential tool for deepening process understanding, which could ultimately reduce the requirement for additional analytical testing [81].

In a second study by Amgen authors, 2D-LC-MS was utilized to identify specific HCPs that co-purify with a panel of 15 different mAbs during protein A chromatography [47]. As a means of normalizing the HCP content, purified mAbs were spiked into reconstituted clarified cell culture fluid (CCCF) from null cells. It was determined that most of the HCP mass present following protein A purification was explained by a set of 14 HCPs common to the 15 mAbs. In addition, the co-purification of an HCP with the drug substance correlated with its overall abundance and avidity for the mAb. Two of the mAbs studied in this work had $(Fab')_2$ regions exhibiting protein A binding. The authors took advantage of this feature to investigate the binding of HCPs to the $(Fab')_2$ region versus the Fc region (common to all of the mAbs in the study). Interestingly, most of the common HCPs associated with both the $(Fab')_2$ and Fc regions of the mAbs. Overall, here again, the analysis of HCPs using 2D-LC-MS during downstream processing continues to improve the understanding of the manufacturing process, and this is expected to simultaneously drive product quality and lower patient risk.

Based on this recent work, the utility of 2D-LC for HCP analysis is clear. However, there remain aspects of the approach that warrant improvement, particularly when it comes to overall analysis time per sample, which can be as long as 10 hours [63]. Indeed, the extra time needed to perform 2D-LC separations has drawn scrutiny and continues to drive investigation of 1D-LC methods with various enhancements [82,83]. For example, Kreimer et al. recently demonstrated a 1D-LC-MS approach, with an analysis time of approximately 2 hours, made possible by the combination of data independent acquisition (DIA) MS and pseudo-data-dependent acquisition (pseudo-DDA) MS [83]. However, this approach did use 2D-LC to create a spectral library of peptides (required for targeted DIA analysis) present in the protein A eluate and required an additional confirmation step to eliminate false-positive matches expected from the data analysis approach. However, continued improvements in speed are expected, especially as techniques such as LC × LC continue to evolve and find use for HCP analysis.

## 2.6 CHARACTERIZATION OF ANTIBODY–DRUG CONJUGATES

ADCs are one of the fastest growing classes of oncology therapeutics. After half a century of research, the approvals of brentuximab vedotin/Adcetris© (2011) trastuzumab emtansine/Kadcyla© (2013) and more recently of inotuzumab ozogamicin/Besponsa© (2017), as well as the re-approval of gemtuzumab ozogamicin/Mylotarg© (2017), have paved the way for ongoing clinical trials that are evaluating more than 80 other ADC candidates [84]. The limited success of first-generation ADCs informed strategies to bring second-generation ADCs to the market, which have higher levels of cytotoxic drug conjugation, lower levels of naked antibodies, and more stable linkers between the drug and the antibody. Furthermore, lessons learned during the past decade are now being used in the development of third-generation molecules [85]. ADCs are more complex than naked mAbs, as the heterogeneity of the conjugates adds to the inherent microvariability of the biomolecules.

The development and optimization of ADCs rely on improving their analytical and bioanalytical characterizations by assessing several CQAs, namely the drug loading distribution (DLD), the amount of naked antibody, the average drug to antibody ratio (DAR), and the residual drug-linker and related product proportions [86].

Figure 2.10 shows the basic structure of a cysteine-linked ADC with the cytotoxic small molecule drug bound to the antibody through a short linker molecule. Figure 2.11 shows that in the case of a cysteine-linked ADC, many different forms

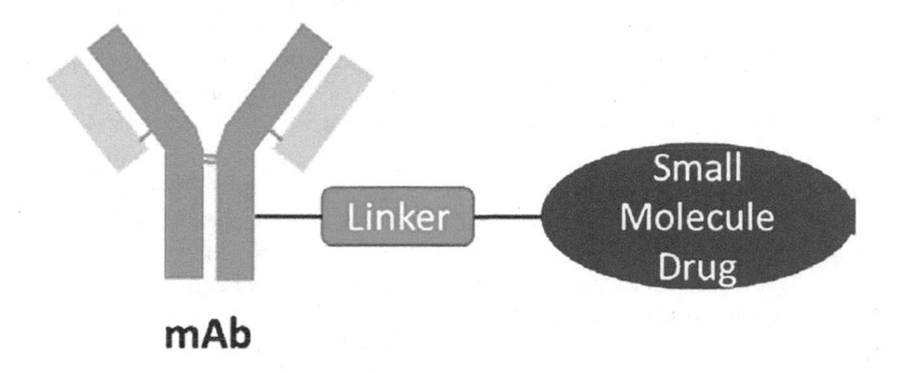


FIGURE 2.10 Cartoon showing the covalent attachment of the cytotoxic small-molecule drug payload to an antibody through a small-molecule linker.

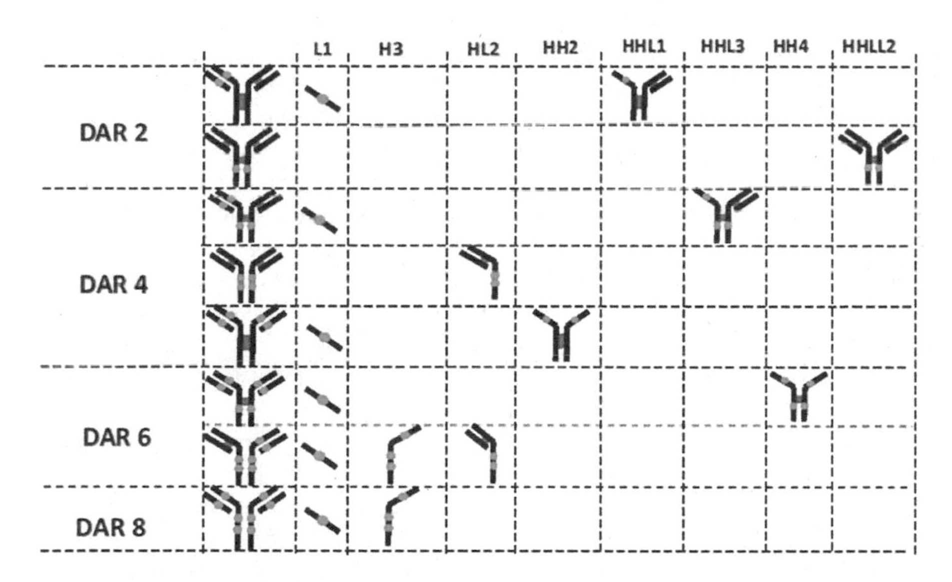


FIGURE 2.11 Drug-loading map showing the possible ways in which a small-molecule drug can be linked to the mAb through cysteine conjugation in the ADC brentuximab vedotin. "L" indicates mAb light chain, and "H" indicates mAb heavy chain. Blue and green dots indicate unconjugated and conjugated cysteines, respectively. Note that several structural isomers can be formed, where the small molecule is localized to different parts of the antibody. (Reprinted from Sarrut, M. et al., *J. Chromatogr. B*, 1032, 91–102, 2016. With permission.)

of the ADC are possible, including some structural isomers, when the conjugation is incomplete. Because of the hybrid nature of ADCs, product quality attributes for both the biological component and the small-molecule components must be considered. Therefore, early developability assessment requires state-of-the-art analytical and structural characterization methods, such as native and ion mobility MS, 2D-LC, and capillary electrophoresis (CE) coupled to MS [87]. These emerging methods provide deep insight into important structural features that are related to ADC function, as well as allow the understanding of ADC biotransformations *in vivo* [88].

Interchain cysteine-conjugated ADCs such as brentuximab vedotin (BV) are manufactured using controlled partial reduction and conjugation chemistry, with drug payloads that typically occur in increments of two (i.e., 0, 2, 4, 6, and 8). Drug loading and distribution can affect the safety and efficacy of the ADC. Conventional 1D-LC coupled with UV absorbance detection, using the HIC mode of separation, can be used to obtain the drug distribution profiles of cysteine-conjugated ADCs. HIC separations are generally incompatible with direct MS detection because of the high levels of non-volatile salt (e.g., >1 M ammonium sulfate) used in the mobile phase. Birdsall et al. reported a heart-cutting 2D-LC method designed for indirectly coupling methods such as HIC to MS detection (LC-LC-QTOF-MS). The resulting method was used for rapid online structural elucidation of species observed in HIC elution profiles of cysteine-conjugated ADCs [54]. The methodology was tested using an IgG1 mAb modified by cysteine conjugation with a non-toxic drug mimic at low, medium, and high levels of conjugation. Identification of HIC peaks based on subunit masses was possible following 2D separation under denaturing RP conditions that promoted dissociation of non-covalently bound mAb subunits. On identification, the average DAR values were determined to be 2.8, 4.4, and 6.0, respectively, in three different ADC batches, based on the relative abundance from the LC-UV data.

In a similar study, Gilroy and Eakin described the development of a suite of heart-cutting 2D-LC methods to characterize peaks eluting from HIC separation of a cysteine-linked ADC, based on an IgG1 and the small-molecule drug auristatin F [40]. In the first of three 2D-LC methods, SEC was used as the $^{2}$D separation to provide a desalting step prior to the determination of intact mass of each HIC peak by native MS. This method clearly established that ADCs containing 2, 4, or 6 linked small-molecule drugs were present in the sample. HIC peaks that likely contained 0 or 8 drugs were observed, but their concentration was low, and the poor MS sensitivity in the SEC mobile phase did not provide definitive mass spectra for these species. In a second method, RP was used as the $^{2}$D separation instead of SEC, again with MS detection following the $^{2}$D separation. The RP mobile phase denatures the protein, resulting in the dissociation of any mAb subunits that are only associated through non-covalent interactions following the reduction of interchain disulfide bonds to form links to the small-molecule drug. This combination of HIC and RP separations then enables localization of the drug loading to different mAb subunits to help understand which of the isomers shown in Figure 2.11 are present. Finally, in a third method, an inline reduction step (using Dithiothreitol (DTT)) and incubation in a large sample loop) was used between the HIC and RP separations to further break any interchain disulfide linkages between mAb subunits. This allowed both further localization of drug loading to light or heavy chains of the mAb.

Whereas the two studies discussed previously involved heart-cutting 2D-LC methods, characterization of cysteine-linked ADCs has also been developed using LC × LC. Such a method involving HIC and RP separations in the first and second dimensions, respectively, and coupled with high-resolution MS detection, was reported by Sarrut et al. and was demonstrated for BV. A first set of experiments was used to optimize the numerous conditions that can affect the quality of the HIC × RP separation [50]. HIC and RP conditions were optimized to prevent salt precipitation owing to solvent mixing during the transfer of fractions from the first to the second dimension and to enhance sensitivity, all while reducing the total analysis time. It was found that adding salt in the sample solvent before HIC injection provided a significant improvement in peak shape in the HIC separation. The gradient profile was also carefully optimized in both dimensions, leading to a two-step gradient in HIC and bracketed gradient in Reversed Phase Liquid Chromatography (RPLC). This study showed that online HIC × RP separation coupled with high-resolution MS detection is a useful method to rapidly (75-minute analysis time) obtain extensive structural information on the peaks observed in the 1D HIC separation, thereby yielding in a single step the precise determination of DLD and the average DAR by HIC. A representative example of this type of separation is shown in Figure 2.12. We see that the arrangement of peaks in the 2D chromatogram is highly ordered, facilitating the identification of the species observed in the sample. It was also demonstrated that the retention data obtained in the first and second dimensions was particularly useful to assist ADC characterization through the identification of subunits

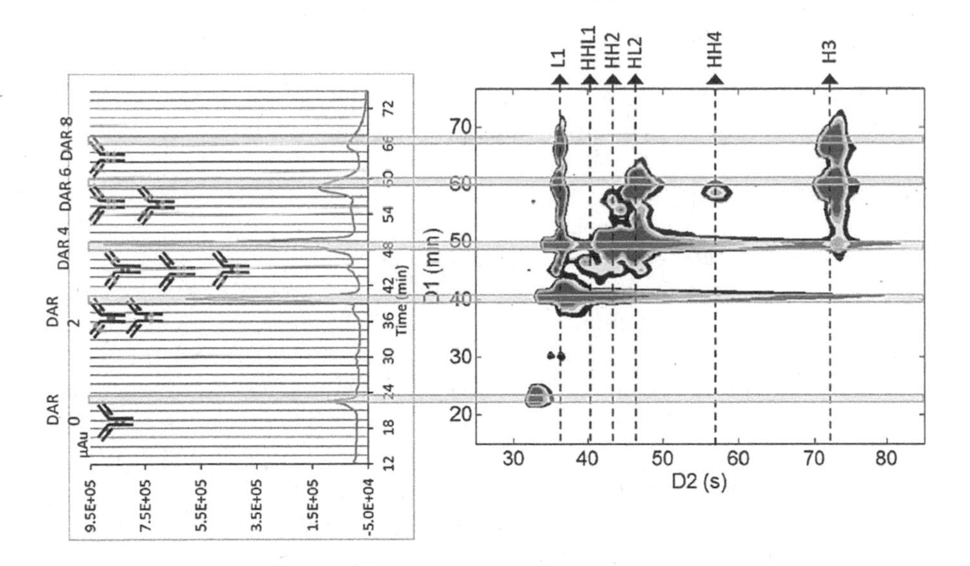


**FIGURE 2.12** Illustration of an LC × LC separation of brentuximab vedotin, using HIC and RP separations in the first and second dimensions, respectively. Proteins elute from the 1D HIC separation (left) in the order of increasing level of small-molecule drug conjugation. In the second dimension, peaks are arranged by protein fragment, leading to a predictable "map" that it is useful for the identification of the species observed. (Reprinted from Sarrut, M. et al., *J. Chromatogr. B*, 1032, 91–102, 2016. With permission.)

after forced degradation studies [89]. Using this methodology, the presence of odd DARs (1, 3, and 5) and their relative abundance were assessed by a systematic evaluation of HIC × RP-UV-MS data for both initial and stressed ADC samples [49].

Interestingly, the versatility of 2D-LC-MS for the characterization of trastuzumab emtansine (T-DM1) has also been demonstrated recently. T-DM1 is generated in a two-step conjugation process. In a first step, mAb lysine residues react with the N-hydroxysuccinimide activated ester of a heterobifunctional linker. Then, DM1, containing a free thiol group, is added in a subsequent step, and it reacts with the maleimide. Under the process conditions, DM1 is conjugated with an average DAR of 3.5. The reaction between DM1 and the linker is not driven to completion; hence, linker without drug can be found to a certain extent [90]. Sandra and coworkers performed mLC-LC separations of T-DM1 by using CEX and RP separations in the first and second dimensions, respectively [48]. Valuable information related to the drug load and its distribution on the protein backbone via lysine linkages was obtained. Subsequently, LC × LC separations with RP columns in both dimensions enabled the determination of small-molecule drug conjugation sites.

## 2.7 ANALYSIS OF RESIDUAL SMALL-MOLECULE DRUGS IN ANTIBODY–DRUG CONJUGATES

As the concept of the ADC is to selectively deliver highly potent cytotoxic small-molecule drugs to the targets via antibodies, the unconjugated free drugs in ADCs can cause off-target toxicity and reduce therapeutic efficacy. The cytotoxic drugs used in ADCs are also called payloads or warheads. So far, the most commonly studied payloads have been auristatins and maytansinoids, for example, in BV and T-DM1. The recently approved Besponsa® and re-approved Mylotarg© use calicheamicin as the payload. There are also many other ADC payloads currently under study in clinical trials [84].

The amount of free small-molecule drug and related impurities present in drug product is a CQA that needs to be accurately and sensitively quantified in ADC products. 1D-LC methods involving RP separations are typically used for small-molecule drug impurity profiling. Chromatographic methods are more accurate, sensitive, and selective compared with other techniques such as ELISA [91,92], but sample pretreatment is required prior to analysis. Common pretreatments include precipitating the proteins and conjugates through the addition of organic solvents [93,94] and using solid-phase extraction (SPE) [93] or restricted-access media [95] to remove protein from the sample. As an alternative to sample pretreatment followed by 1D-LC, 2D-LC methods can be used for the determination of free drug and impurities in an ADC sample, with no or minimum sample pretreatment. This can be a significant advantage, as the workflow can be automated, resulting in higher sample throughput and better sensitivity, in addition to providing comprehensive information of other quality attributes of the sample besides free drug assay [58,96,97].

2D-LC methods provide opportunities to effectively integrate in a single analysis methods that have historically been used separately to address the large- and small-molecule components of ADC samples. Using an SEC method in the first dimension enables the separation of protein and small-molecule species into two distinct elution fractions. The late-eluting small-molecule peak can be transferred to the second dimension through a heart-cutting approach, and the different small-molecule species in the sample can then be separated using an RP column. Such an online heart-cutting SEC-RP 2D-LC UV/MS method was reported for the assay of unconjugated free drug and related impurities in ADC product and stability study samples [58]. As shown in Figure 2.13, the ADC samples were injected directly onto an SEC column in the first dimension, which separated ADC monomer, dimer, and other high-molecular-weight species (HMWS), and the small-molecule species coeluted as one single peak. The small-molecule peak was transferred to the second dimension, and an RP column was employed to further resolve the peak into free drug, linker-drug, N-acetylcysteine (NAC) adduct (a process-related impurity), and degradation products from ADC stability samples. MS was used to identify the impurities and degradation products. UV was used to quantify the small-molecule species, and good method precision, accuracy, linearity, and sensitivity were demonstrated by method validation. This method is 10-fold more sensitive than a 1D-LC analysis using the

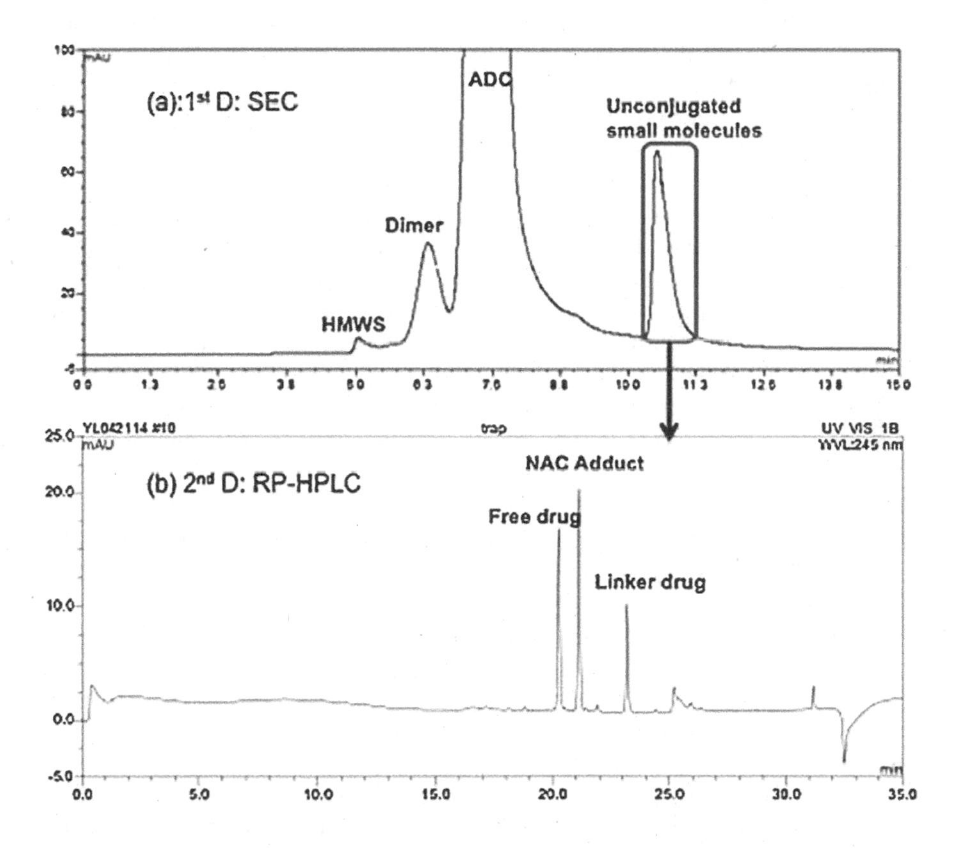


**FIGURE 2.13** Analysis of free small molecules in ADC samples by 2D-LC. (a) SEC separation in the first dimension that separates the unconjugated small molecules from ADC; (b) RPLC separation in the second dimension that separates the unconjugated small molecules. (Reprinted from Li, Y. et al., *J. Chromatogr. A*, 1393, 81–88, 2015. With permission.)

same RP method after protein has been precipitated from the sample. A similar SEC-RP 2D-LC approach was used to investigate a recovery issue during the validation of a precipitation method [51].

Online mixed-mode SPE-RPLC-MS was reported for sensitive quantification of free drug species in antibody-fluorophore conjugate (AFC), an ADC surrogate to brentuximab vedotin [96]. The ADC was eluted quickly from a mixed-mode AEX SPE column in the first dimension, whereas the free drug species were retained on the SPE column. The small-molecule peak was then eluted and transferred to the second dimension and further separated using a superficially porous C18 RP column. Assay sensitivity was found to be two orders of magnitude better by using MS detection in comparison with UV absorbance detection, with a nominal limit of quantitation of 0.30 ng/mL. Another online SPE-RPLC-MS/MS method was reported to sensitively and selectively quantify the free drug DM1 in human serum after the intravenous administration of an investigational ADC containing DM1 as the payload [52]. The reported Lower Limit of Quantitation (LLOQ) was 200 pg/mL, using just 25 μL of human serum.

Before conjugating the small-molecule drug to the antibody, a linker-drug intermediate is synthesized (see Figure 2.10). Characterization of the impurity profile of this linker-drug intermediate is essential to understand the conjugation process and to control the purity of downstream conjugates. The linker-drug intermediates are typically complex and have many impurities owing to the reactive nature of the linker. The combination of comprehensive and heart-cutting 2D-LC was reported to profile an antibody–antibiotic conjugate (AAC) drug linker [98]. First, comprehensive 2D-LC was employed to give a quick bird's-eye view of the overall impurity profile, and then, multiple heart-cutting 2D-LC was used to selectively quantify the co-eluted peaks of interest. Recent developments in instrumentation and software for 2D-LC not only enable the use of robust methods in routine quality control (QC) analysis [99] but also provide the analysis with more options to use different 2D-LC approaches, depending on the goals of the analysis. For example, the high-resolution sampling interface for 2D-LC (i.e., sLC × LC) has been used for qualitative and quantitative analyses of complex ADC linker-drug intermediates [100]. In this case, the Limit of Quantitation (LOQ) was reported to be less than 0.01%, which demonstrated that the 2D-LC method was over five-fold more sensitive than a comparable 1D-LC separation used for this purpose.

## 2.8 CHARACTERIZATION OF FORMULATION INGREDIENTS IN PROTEIN PRODUCTS

As the vast majority of drugs are administered to patients in formulated products, the impact of formulation ingredients on drugs and on patients should not be underestimated [101]. The typical formulation excipients used in therapeutic protein products are polysorbates, sucrose, salts, histidine, etc., to stabilize proteins, control pH, adjust tonicity, and provide reproducible biological activities. The distinct differences between these ingredients themselves, and relative to proteins, in terms of molecular size, heterogeneity, charge, polarity, and hydrophobicity present many challenges to analysis using conventional 1D-LC-based methods.

The development of 2D-LC in recent years has enabled the analysis of complex drug formulation samples and has provided insights that were not attainable previously [1,97,102]. The majority of therapeutic protein products contain surfactants such as polysorbate to stabilize proteins and prevent their aggregation and precipitation [103]. Polysorbates are highly heterogeneous as a result of both the manufacturing process and degradation pathways. Although many efforts have been reported to characterize the heterogeneity of polysorbate standards and the stability of polysorbate in the absence of proteins [104–107], it has been extremely difficult to study the stability of polysorbate in protein formulations, owing to the strong interference from proteins. Multi-dimensional LC coupled with charged aerosol detection (CAD) and MS has been demonstrated as a powerful tool to overcome these challenges [60]. A mixed-mode column with both AEX and RP characteristics was used in the first dimension with an acidic mobile phase to elute the positively charged proteins while retaining the neutral polysorbate esters based on hydrophobic interaction, as show in Figure 2.14. The polysorbate esters were eluted by increasing the mobile-phase organic solvent percentage, transferred to the second dimension by heart-cutting, and separated using a high-performing RP method. Individual peaks were identified by $MS^n$ as mono- and diesters of poly(oxyethylene) (POE) sorbitan, POE isosorbide, and POE esters with different chain lengths of the fatty acid hydrophobic tails. Another 2D-LC method involving CEX and RP separations in the first and second dimensions was developed in the same study [60] to profile the polyol degradation products resulting from the polar head groups of polysorbate. It was found

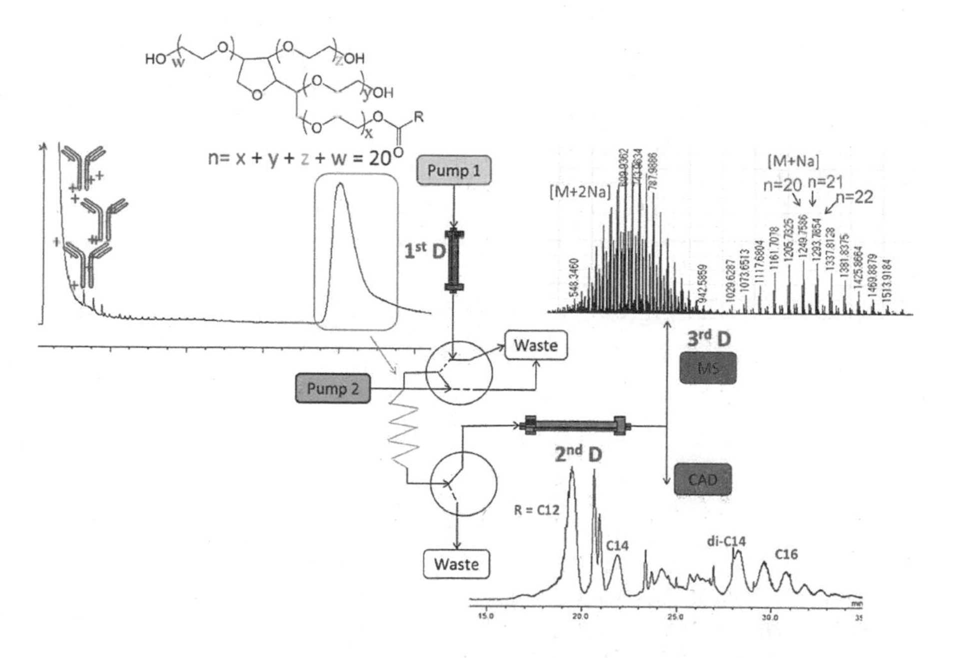


**FIGURE 2.14** Characterization of molecular heterogeneity of polysorbate 20 in a mAb drug product by 2D-LC/MS. (Reprinted from Li, Y.et al., *Anal. Chem.* 86, 5150–5157, 2014. With permission.)

that different polysorbate esters degrade at different rates, and some esters exhibit different degradation behavior in the presence of protein compared with their behavior in formulations without protein. This further emphasized that it is critical to study the molecular heterogeneity and stability of polysorbate in the presence of protein instead of studying the stability of polysorbate standards.

As discussed earlier, the formulation ingredients in mAb drug products vary widely in terms of molecular size, charge, polarity, and hydrophobicity. As a result, mixed-mode and Hydrophilic Interaction Liquid Chromatography (HILIC) separation modes are frequently used to characterize these formulation ingredients. Mixed-mode columns can add dimensionality to a separation by using a single column with a multi-phase mobile-phase elution program that leverages a particular mode of interaction during a particular elution phase [108]. The example described previously involving the study of polysorbate also demonstrated this type of approach [60]. Online coupling of an SEC column with a mixed-mode column has been described for the purpose of profiling biopharmaceutical drug formulations [64]. In this study, proteins and excipients were separated by an SEC column, and then, a column switching valve was used to transfer the later eluted excipient peak into the a mixed-mode column to separate different excipients, including cations $Na^+$ and $K^+$, the anion $Cl^-$, the non-ionic hydrophobic surfactant polysorbate 80, and the hydrophilic sugar sucrose. Multiple applications of the method were demonstrated through the analysis of mAb, ADC, and vaccine drug product samples.

A heart-cutting SEC-HILIC 2D-LC method was reported to characterize and quantify histidine degradation in protein formulation samples [59]. The SEC column was used to remove protein matrix interference in the first dimension. Histidine and its degradant were analyzed in the second dimension by using a ZIC-HILIC column. A small transfer volume and fast gradient at high flow rate in the second dimension were used in order to mitigate the solvent mismatch between the SEC and HILIC mobile phases. The histidine degradant was determined to be trans-urocanic acid and was quantified by the combination of this 2D-LC method with stable-isotope labeling MS detection. It was also found the $^{1}$D diode array detector was causing the degradation of photolabile trans-urocanic acid to form cis-urocanic acid.

## 2.9 USE OF TWO-DIMENSIONAL LIQUID CHROMATOGRAPHY IN QUALITY CONTROL

An important consideration for the future use of 2D-LC will be the scope of implementation of 2D-LC methods. It is clear from the work discussed throughout this chapter and in recent years that 2D-LC is becoming established as a powerful tool for rapid and deep characterization of mAbs and related materials in research and development environments. Some very different but equally important environments where 2D-LC may be used are in-process control (IPC), QC, and quality assurance (QA). Only time will tell about the extent to which 2D-LC methods are adopted in QC and QA. However, Largy and coworkers recently indicated that they perceive several advantages of 2D-LC over conventional 1D-LC for the analysis of mAbs and ADCs in regulated environments [45]. These include the ability to establish a robust, generic 2D separation approach that accommodates a range of possible $^{1}$D separations, including CEX, HIC, and SEC, while coupling indirectly to MS detection through a $^{2}$D RP separation. They stated that the online 2D separation approach is less time- and resource-consuming compared with the offline 2D separation approaches that have been used historically to obtain MS data for MS-incompatible separations. A current trend is favoring the development of 2D-LC methodologies that enable the augmentation of existing 1D-LC separations with additional dimensions of separation (second and higher), in contrast to developing multi-dimensional methods from the ground up. An example of this is demonstrated in the recent work of Gstöttner and coworkers [39]. This saves time in method development and increases the likelihood of successful method transfer to other laboratories.

Finally, we would like to highlight the recent wide-ranging paper of Sandra and coworkers that demonstrates the versatility of the state-of-the-art 2D separations for the characterization of mAbs [42]. Several aspects of the paper have been touched on already in this section. However, we emphasize here that the work very nicely shows by the way of comparison, using a single set of samples, how different implementations of 2D-LC can complement each other in the characterization of mAbs. This is both in terms of the mode of 2D separation, ranging from simple single heart-cut operation to fully comprehensive 2D separations, and in terms of the modes of constituent 1D separations, including affinity (protein A), CEX, SEC, and RP.

## 2.10 FUTURE OUTLOOK

As reviewed here, 2D-LC separations with MS and other detection methods clearly facilitate a deep structural understanding of large molecules such as mAbs and ADCs. The additional chromatographic selectivity and resolution of 2D-LC compared with conventional 1D-LC methods enable the direct and efficient identification of different species present in these materials. This capability will be critically important for the characterization of drug products involving both the drug substance and excipients, studies of drug distribution and metabolism, and support of manufacturing processes and product release. Future development of 2D-LC methods will likely involve the exploration of the full array of chromatographic techniques already developed for the characterization of antibodies by conventional 1D-LC (e.g., SEC, IEX, HIC, and HILIC). While the choice of separation mechanism for use in the first dimension is quite flexible, RPLC will be most frequently used in the second dimension, as it is directly compatible with MS detection. Recent applications of such approaches have involved SEC × RP-MS, CEX × RP-MS, and HIC × RP-MS. Moreover, recent improvements made to specific separation modes such as SEC [109], CEX [110], RP [111], HIC [112], and HILIC [113,114] separations will undoubtedly further improve the performance of 2D-LC–MS analyses of protein materials. In very recent work, multiple groups have begun to explore ways to further improve the power of 2D-LC separations. Ehkirch, Cianférani, and coworkers developed an LC × LC separation involving HIC and SEC separations in the first and second dimensions, followed by Ion Mobility Spectrometry (IMS)-MS detection [38]. This separation, which is described as involving a total of four dimensions, was shown to be very useful for deep and efficient characterization of ADCs. In a different study, Gstöttner, Kopf, and coworkers described a 2D-LC system with inline protein reduction and proteolytic digestion between the first and second dimensions of separation [39]. This system was applied to the detailed identification of microvariants of IgGs. All of these developments will become increasingly important, as the complexity of these materials continues to increase, as in the cases of heterogeneous mAbs, ADCs, and related products such as Fc-fusion proteins and peptides, Fab fragments, bi- (bsAbs and bsADCs) and multi-specific antibodies, and recombinant and polyclonal antibodies (pAbs) [38].

## REFERENCES

1. D. Stoll, J. Danforth, K. Zhang, A. Beck, Characterization of therapeutic antibodies and related products by two-dimensional LC coupled with UV absorbance and mass spectrometric detection, *Journal of Chromatography B*. 1032 (2016) 51–60. doi:10.1016/j.jchromb.2016.05.029.
2. A. Beck, T. Wurch, C. Bailly, N. Corvaia, Strategies and challenges for the next generation of therapeutic antibodies, *Nature Reviews Immunology*. 10 (2010) 345–352. doi:10.1038/nri2747.
3. H. Kaplon, J.M. Reichert, Antibodies to watch in 2018, *MAbs*. (2018). doi:10.1080/19420862.2018.1415671.
4. P.J. Carter, G.A. Lazar, Next generation antibody drugs: Pursuit of the "high-hanging fruit," *Nature Reviews Drug Discovery*. (2017). doi:10.1038/nrd.2017.227.
5. J. Fang, C. Doneanu, W.R. Alley, Y.Q. Yu, A. Beck, W. Chen, Advanced assessment of the physicochemical characteristics of Remicade® and Inflectra® by sensitive LC/MS techniques, *MAbs*. 8 (2016) 1021–1034. doi:10.1080/19420862.2016.1193661.
6. A. Beck, E. Wagner-Rousset, D. Ayoub, A. Van Dorsselaer, S. Sanglier-Cianférani, Characterization of therapeutic antibodies and related products, *Analytical Chemistry*. 85 (2013) 715–736. doi:10.1021/ac3032355.
7. J. Giorgetti, V. D'Atri, J. Canonge, A. Lechner, D. Guillarme, O. Colas, E. Wagner-Rousset, A. Beck, E. Leize-Wagner, Y.-N. François, Monoclonal antibody N-glycosylation profiling using capillary electrophoresis—Mass spectrometry: Assessment and method validation, *Talanta*. 178 (2018) 530–537. doi:10.1016/j.talanta.2017.09.083.
8. E. Largy, F. Cantais, G. Van Vyncht, A. Beck, A. Delobel, Orthogonal liquid chromatography—Mass spectrometry methods for the comprehensive characterization of therapeutic glycoproteins, from released glycans to intact protein level, *Journal of Chromatography A*. 1498 (2017) 128–146. doi:10.1016/j.chroma.2017.02.072.
9. E. Wagner-Rousset, S. Fekete, L. Morel-Chevillet, O. Colas, N. Corvaïa, S. Cianférani, D. Guillarme, A. Beck, Development of a fast workflow to screen the charge variants of therapeutic antibodies, *Journal of Chromatography A*. 1498 (2017) 147–154. doi:10.1016/j.chroma.2017.02.065.
10. D.R. Stoll, T.D. Maloney, Recent advances in two-dimensional liquid chromatography for pharmaceutical and biopharmaceutical analysis, *LCGC North America*. 35 (2017) 680–687.
11. D.R. Stoll, P.W. Carr, Two-dimensional liquid chromatography: A state of the art tutorial, *Analytical Chemistry*. 89 (2017) 519–531. doi:10.1021/acs.analchem.6b03506.
12. B.W.J. Pirok, A.F.G. Gargano, P.J. Schoenmakers, Optimizing separations in on-line comprehensive two-dimensional liquid chromatography, *Journal of Separation Science*. (2017). doi:10.1002/jssc.201700863.
13. D. Stoll, Introduction to two-dimensional liquid chromatography—Theory and practice, in: M. Holcapek, W.C. Byrdwell (Eds.), *Handbook of Advanced Chromatography/Mass Spectrometry Techniques*, Elsevier, London, UK, 2017: pp. 227–286.

14. D. Li, C. Jakob, O. Schmitz, Practical considerations in comprehensive two-dimensional liquid chromatography systems (LC × LC) with reversed-phases in both dimensions, *Analytical and Bioanalytical Chemistry*. 407 (2014) 153–167. doi:10.1007/s00216-014-8179-8.
15. P.J. Marriott, P.J. Schoenmakers, Z. Wu, Nomenclature and conventions in comprehensive multidimensional chromatography—An update, *LC-GC Europe*. 25 (2012) 266, 268, 270, 272–275.
16. S.W. Simpkins, J.W. Bedard, S.R. Groskreutz, M.M. Swenson, T.E. Liskutin, D.R. Stoll, Targeted three-dimensional liquid chromatography: A versatile tool for quantitative trace analysis in complex matrices, *Journal of Chromatography A*. 1217 (2010) 7648–7660. doi:10.1016/j.chroma.2010.09.023.
17. K. Zhang, Y. Li, M. Tsang, N.P. Chetwyn, Analysis of pharmaceutical impurities using multi-heartcutting 2D LC coupled with UV-charged aerosol MS detection: Liquid chromatography, *Journal of Separation Science*. 36 (2013) 2986–2992. doi:10.1002/jssc.201300493.
18. M. Pursch, S. Buckenmaier, Loop-based multiple heart-cutting two-dimensional liquid chromatography for target analysis in complex matrices, *Analytical Chemistry*. 87 (2015) 5310–5317. doi:10.1021/acs.analchem.5b00492.
19. S.R. Groskreutz, M.M. Swenson, L.B. Secor, D.R. Stoll, Selective comprehensive multi-dimensional separation for resolution enhancement in high performance liquid chromatography, Part I—Principles and instrumentation, *Journal of Chromatography A*. 1228 (2012) 31–40. doi:10.1016/j.chroma.2011.06.035.
20. S.R. Groskreutz, M.M. Swenson, L.B. Secor, D.R. Stoll, Selective comprehensive multi-dimensional separation for resolution enhancement in high performance liquid chromatography, Part II—Applications, *Journal of Chromatography A*. 1228 (2012) 41–50. doi:10.1016/j.chroma.2011.06.038.
21. I. Francois, K. Sandra, P. Sandra. History, evolution and optimization aspects of comprehensive two-dimensional liquid chromatography. In L. Mondello, (Ed.), *Comprehensive Chromatography in Combination with Mass Spectrometry*, 1st edn, John Wiley & Sons, New York, 2011, 281–330.
22. D.R. Stoll, K. Shoykhet, P. Petersson, S. Buckenmaier, Active Solvent Modulation—A valve-based approach to improve separation compatibility in two-dimensional liquid chromatography, *Analytical Chemistry*. 89 (2017) 9260–9267. doi:10.1021/acs.analchem.7b02046.
23. D.R. Stoll, D.C. Harmes, J. Danforth, E. Wagner-Rousset, D. Guillarme, S. Fekete, A. Beck, Direct identification of rituximab main isoforms and subunit analysis by online selective comprehensive two-dimensional liquid chromatography—Mass spectrometry, *Analytical Chemistry*. 87 (2015) 8307–8315. doi:10.1021/acs.analchem.5b01578.
24. M. Egeness, M. Breadmore, E. Hilder, R.A. Shellie, The modulator in comprehensive two-dimensional liquid chromatography, *LCGC Europe*. 5 (2016) 268–276.
25. A. Schellinger, D. Stoll, P. Carr, High speed gradient elution reversed-phase liquid chromatography, *Journal of Chromatography A*. 1064 (2005) 143–156. doi:10.1016/j.chroma.2004.12.017.
26. A.P. Schellinger, D.R. Stoll, P.W. Carr, High speed gradient elution reversed phase liquid chromatography of bases in buffered eluents—II. Full equilibrium, *Journal of Chromatography A*. 1192 (2008) 54–61. doi:10.1016/j.chroma.2008.02.049.
27. A.P. Schellinger, D.R. Stoll, P.W. Carr, High-speed gradient elution reversed-phase liquid chromatography of bases in buffered eluents—I. Retention repeatability and column re-equilibration, *Journal of Chromatography A*. 1192 (2008) 41–53. doi:10.1016/j.chroma.2008.01.062.
28. R.E. Murphy, M.R. Schure, J.P. Foley, Effect of sampling rate on resolution in comprehensive two-dimensional liquid chromatography, *Analytical Chemistry*. 70 (1998) 1585–1594. doi:10.1021/ac971184b.

29. J.M. Davis, D.R. Stoll, P.W. Carr, Effect of first-dimension undersampling on effective peak capacity in comprehensive two-dimensional separations, *Analytical Chemistry.* 80 (2008) 461–473. doi:10.1021/ac071504j.
30. F. Bedani, P.J. Schoenmakers, H.-G. Janssen, Theories to support method development in comprehensive two-dimensional liquid chromatography—A review, *Journal of Separation Science.* 35 (2012) 1697–1711. doi:10.1002/jssc.201200070.
31. L.W. Potts, D.R. Stoll, X. Li, P.W. Carr, The impact of sampling time on peak capacity and analysis speed in on-line comprehensive two-dimensional liquid chromatography, *Journal of Chromatography A.* 1217 (2010) 5700–5709. doi:10.1016/j.chroma.2010.07.009.
32. Y. Huang, H. Gu, M. Filgueira, P.W. Carr, An experimental study of sampling time effects on the resolving power of on-line two-dimensional high performance liquid chromatography, *Journal of Chromatography A.* 1218 (2011) 2984–2994. doi:10.1016/j.chroma.2011.03.032.
33. D.R. Stoll, X. Wang, P.W. Carr, Comparison of the practical resolving power of one- and two-dimensional high-performance liquid chromatography analysis of metabolomic samples, *Analytical Chemistry.* 80 (2008) 268–278. doi:10.1021/ac701676b.
34. M. Sarrut, F. Rouvière, S. Heinisch, Theoretical and experimental comparison of one dimensional versus on-line comprehensive two dimensional liquid chromatography for optimized sub-hour separations of complex peptide samples, *Journal of Chromatography A.* 1498 (2017) 183–195. doi:10.1016/j.chroma.2017.01.054.
35. B.W.J. Pirok, S. Pous-Torres, C. Ortiz-Bolsico, G. Vivó-Truyols, P.J. Schoenmakers, Program for the interpretive optimization of two-dimensional resolution, *Journal of Chromatography A.* 1450 (2016) 29–37. doi:10.1016/j.chroma.2016.04.061.
36. G. Vivó-Truyols, S. van der Wal, P.J. Schoenmakers, Comprehensive study on the optimization of online two-dimensional liquid chromatographic systems considering losses in theoretical peak capacity in first- and second-dimensions: A Pareto-optimality approach, *Analytical Chemistry.* 82 (2010) 8525–8536. doi:10.1021/ac101420f.
37. M. Sarrut, A. D'Attoma, S. Heinisch, Optimization of conditions in on-line comprehensive two-dimensional reversed phase liquid chromatography. Experimental comparison with one-dimensional reversed phase liquid chromatography for the separation of peptides, *Journal of Chromatography A.* 1421 (2015) 48–59. doi:10.1016/j.chroma.2015.08.052.
38. A. Ehkirch, V. D'Atri, F. Rouvière, O. Hernandez-Alba, A. Goyon, O. Colas, M. Sarrut, A. Beck, D. Guillarme, S. Heinisch, S. Cianférani, An online four-dimensional HICxSEC-IMxMS methodology for proof-of-concept of antibody drug conjugates characterization, *Analytical Chemistry.* 90 (2018) 1578–1586.
39. C.J. Gstöttner, D. Klemm, M. Haberger, A. Bathke, H. Wegele, C.H. Bell, R. Kopf, Fast and automated characterization of antibody variants with 4D-HPLC/MS, *Analytical Chemistry.* 90 (2018) 2119–2125. doi:10.1021/acs.analchem.7b04372.
40. J.J. Gilroy, C.M. Eakin, Characterization of drug load variants in a thiol linked antibody-drug conjugate using multidimensional chromatography, *Journal of Chromatography B.* 1060 (2017) 182–189. doi:10.1016/j.jchromb.2017.06.005.
41. A. Williams, E.K. Read, C.D. Agarabi, S. Lute, K.A. Brorson, Automated 2D-HPLC method for characterization of protein aggregation with in-line fraction collection device, *Journal of Chromatography B.* 1046 (2017) 122–130. doi:10.1016/j.jchromb.2017.01.021.
42. K. Sandra, M. Steenbeke, I. Vandenheede, G. Vanhoenacker, P. Sandra, The versatility of heart-cutting and comprehensive two-dimensional liquid chromatography in monoclonal antibody clone selection, *Journal of Chromatography A.* 1523 (2017) 283–292. doi:10.1016/j.chroma.2017.06.052.

43. A. Yan, V. Shikha, Y. Chen, S. Yu, Y. Zhang, S. Kelner, S. Mengisen, D. Richardson, Z. Chen, Forced degradation study of monoclonal antibody using two- dimensional liquid chromatography, *Journal of Chromatography & Separation Techniques*. 08 (2017) 2. doi:10.4172/2157-7064.1000365.
44. H. Luo, W. Zhong, J. Yang, P. Zhuang, F. Meng, J. Caldwell, B. Mao, C.J. Welch, 2D-LC as an on-line desalting tool allowing peptide identification directly from MS unfriendly HPLC methods, *Journal of Pharmaceutical and Biomedical Analysis*. 137 (2017) 139–145. doi:10.1016/j.jpba.2016.11.012.
45. E. Largy, A. Catrain, G. Van Vyncht, A. Delobel, 2D-LC–MS for the analysis of monoclonal antibodies and antibody–drug conjugates in a regulated environment, *Current Trends in Mass Spectrometry*. 14 (2016) 29–35.
46. M. Sorensen, D.C. Harmes, D.R. Stoll, G.O. Staples, S. Fekete, D. Guillarme, A. Beck, Comparison of originator and biosimilar therapeutic monoclonal antibodies using comprehensive two-dimensional liquid chromatography coupled with time-of-flight mass spectrometry, *MAbs*. 8 (2016) 1224–1234. doi:10.1080/19420862.2016.1203497.
47. Q. Zhang, A.M. Goetze, H. Cui, J. Wylie, B. Tillotson, A. Hewig, M.P. Hall, G.C. Flynn, Characterization of the co-elution of host cell proteins with monoclonal antibodies during protein A purification, *Biotechnology Progress*. 32 (2016) 708–17. doi:10.1002/btpr.2272.
48. K. Sandra, G. Vanhoenacker, I. Vandenheede, M. Steenbeke, M. Joseph, P. Sandra, Multiple heart-cutting and comprehensive two-dimensional liquid chromatography hyphenated to mass spectrometry for the characterization of the antibody-drug conjugate ado-trastuzumab emtansine, *Journal of Chromatography B*. 1032 (2016) 119–130. doi:10.1016/j.jchromb.2016.04.040.
49. M. Sarrut, A. Corgier, S. Fekete, D. Guillarme, D. Lascoux, M.-C. Janin-Bussat, A. Beck, S. Heinisch, Analysis of antibody-drug conjugates by comprehensive on-line two-dimensional hydrophobic interaction chromatography x reversed phase liquid chromatography hyphenated to high resolution mass spectrometry. I—Optimization of separation conditions, *Journal of Chromatography B*. 1032 (2016) 103–111. doi:10.1016/j.jchromb.2016.06.048.
50. M. Sarrut, S. Fekete, M.-C. Janin-Bussat, O. Colas, D. Guillarme, A. Beck, S. Heinisch, Analysis of antibody-drug conjugates by comprehensive on-line two-dimensional hydrophobic interaction chromatography x reversed phase liquid chromatography hyphenated to high resolution mass spectrometry. II- Identification of sub-units for the characterization of even and odd load drug species, *Journal of Chromatography B*. 1032 (2016) 91–102. doi:10.1016/j.jchromb.2016.06.049.
51. Y. Li, C. Stella, L. Zheng, C. Bechtel, J. Gruenhagen, F. Jacobson, C.D. Medley, Investigation of low recovery in the free drug assay for antibody drug conjugates by size exclusion—Reversed phase two dimensional-liquid chromatography, *Journal of Chromatography B*. 1032 (2016) 112–118. doi:10.1016/j.jchromb.2016.05.011.
52. O. Heudi, S. Barteau, F. Picard, O. Kretz, Quantitative analysis of maytansinoid (DM1) in human serum by on-line solid phase extraction coupled with liquid chromatography tandem mass spectrometry—Method validation and its application to clinical samples, *Journal of Pharmaceutical and Biomedical Analysis*. 120 (2016) 322–332. doi:10.1016/j.jpba.2015.12.026.
53. A. Farrell, S. Mittermayr, B. Morrissey, N. Mc Loughlin, N. Navas Iglesias, I.W. Marison, J. Bones, Quantitative host cell protein analysis using two dimensional data independent LC–MS$^E$, *Analytical Chemistry*. 87 (2015) 9186–9193. doi:10.1021/acs.analchem.5b01377.

54. R.E. Birdsall, H. Shion, F.W. Kotch, A. Xu, T.J. Porter, W. Chen, A rapid on-line method for mass spectrometric confirmation of a cysteine-conjugated antibody-drug-conjugate structure using multidimensional chromatography, *MAbs.* 7 (2015) 1036–1044. doi:10.1080/19420862.2015.1083665.
55. C.E. Doneanu, M. Anderson, B.J. Williams, M.A. Lauber, A. Chakraborty, W. Chen, Enhanced detection of low-abundance host cell protein impurities in high-purity monoclonal antibodies down to 1 ppm using ion mobility mass spectrometry coupled with multidimensional liquid chromatography, *Analytical Chemistry.* (2015) 10283–10291. doi:10.1021/acs.analchem.5b02103.
56. Y. Shen, G. Zhang, J. Yang, Y. Qiu, T. McCauley, L. Pan, J. Wu, Online 2D-LC-MS/MS assay to quantify therapeutic protein in human serum in the presence of pre-existing antidrug antibodies, *Analytical Chemistry.* 87 (2015) 8555–8563. doi:10.1021/acs.analchem.5b02293.
57. G. Vanhoenacker, I. Vandenheede, F. David, P. Sandra, K. Sandra, Comprehensive two-dimensional liquid chromatography of therapeutic monoclonal antibody digests, *Analytical and Bioanalytical Chemistry.* 407 (2015) 355–366. doi:10.1007/s00216-014-8299-1.
58. Y. Li, C. Gu, J. Gruenhagen, K. Zhang, P. Yehl, N.P. Chetwyn, C.D. Medley, A size exclusion-reversed phase two dimensional-liquid chromatography methodology for stability and small molecule related species in antibody drug conjugates, *Journal of Chromatography A.* 1393 (2015) 81–88. doi:10.1016/j.chroma.2015.03.027.
59. C. Wang, S. Chen, J.A. Brailsford, A.P. Yamniuk, A.A. Tymiak, Y. Zhang, Characterization and quantification of histidine degradation in therapeutic protein formulations by size exclusion-hydrophilic interaction two dimensional-liquid chromatography with stable-isotope labeling mass spectrometry, *Journal of Chromatography A.* 1426 (2015) 133–139. doi:10.1016/j.chroma.2015.11.065.
60. Y. Li, D. Hewitt, Y. Lentz, J. Ji, T. Zhang, K. Zhang, Characterization and stability study of polysorbate 20 in therapeutic monoclonal antibody formulation by multidimensional ultrahigh-performance liquid chromatography—Charged aerosol detection—Mass spectrometry, *Analytical Chemistry.* 86 (2014) 5150–5157. doi:10.1021/ac5009628.
61. M.M. St. Amand, B.A. Ogunnaike, A.S. Robinson, Development of at-line assay to monitor charge variants of MAbs during production, *Biotechnology Progress.* 30 (2014) 249–255. doi:10.1002/btpr.1848.
62. Q. Zhang, A.M. Goetze, H. Cui, J. Wylie, S. Trimble, A. Hewig, G.C. Flynn, Comprehensive tracking of host cell proteins during monoclonal antibody purifications using mass spectrometry, *MAbs.* 6 (2014) 659–670. doi:10.4161/mabs.28120.
63. C. Doneanu, A. Xenopoulos, K. Fadgen, J. Murphy, S.J. Skilton, H. Prentice, M. Stapels, W. Chen, Analysis of host-cell proteins in biotherapeutic proteins by comprehensive online two-dimensional liquid chromatography/mass spectrometry, *MAbs.* 4 (2012) 24–44. doi:10.4161/mabs.4.1.18748.
64. Y. He, O.V. Friese, M.R. Schlittler, Q. Wang, X. Yang, L.A. Bass, M.T. Jones, On-line coupling of size exclusion chromatography with mixed-mode liquid chromatography for comprehensive profiling of biopharmaceutical drug product, *Journal of Chromatography A.* 1262 (2012) 122–129. doi:10.1016/j.chroma.2012.09.012.
65. M.T. Mazur, R.S. Seipert, D. Mahon, Q. Zhou, T. Liu, A platform for characterizing therapeutic monoclonal antibody breakdown products by 2D chromatography and top-down mass spectrometry, *The AAPS Journal.* 14 (2012) 530–541. doi:10.1208/s12248-012-9361-6.
66. M. Alvarez, G. Tremintin, J. Wang, M. Eng, Y.-H. Kao, J. Jeong, V.T. Ling, O.V. Borisov, On-line characterization of monoclonal antibody variants by liquid chromatography–mass spectrometry operating in a two-dimensional format, *Analytical Biochemistry.* 419 (2011) 17–25. doi:10.1016/j.ab.2011.07.033.

67. R.J. Harris, B. Kabakoff, F.D. Macchi, F.J. Shen, M. Kwong, J.D. Andya, S.J. Shire, N. Bjork, K. Totpal, A.B. Chen, Identification of multiple sources of charge heterogeneity in a recombinant antibody, *Journal of Chromatography B: Biomedical Sciences and Applications*. 752 (2001) 233–245. doi:10.1016/S0378-4347(00)00548-X.
68. S. Fekete, A. Beck, J. Fekete, D. Guillarme, Method development for the separation of monoclonal antibody charge variants in cation exchange chromatography, Part II: pH gradient approach, *Journal of Pharmaceutical and Biomedical Analysis*. 102 (2015) 282–289. doi:10.1016/j.jpba.2014.09.032.
69. S. Fekete, A. Beck, J. Fekete, D. Guillarme, Method development for the separation of monoclonal antibody charge variants in cation exchange chromatography, Part I: Salt gradient approach, *Journal of Pharmaceutical and Biomedical Analysis*. 102 (2015) 33–44. doi:10.1016/j.jpba.2014.08.035.
70. R.S. Rogers, N.S. Nightlinger, B. Livingston, P. Campbell, R. Bailey, A. Balland, Development of a quantitative mass spectrometry multi-attribute method for characterization, quality control testing and disposition of biologics, *MAbs*. 7 (2015) 881–890. doi:10.1080/19420862.2015.1069454.
71. R.S. Rogers, M. Abernathy, D.D. Richardson, J.C. Rouse, J.B. Sperry, P. Swann, J. Wypych, C. Yu, L. Zang, R. Deshpande, A view on the importance of "Multi-Attribute Method" for measuring purity of biopharmaceuticals and improving overall control strategy, *The AAPS Journal*. 20 (2018). doi:10.1208/s12248-017-0168-3.
72. Q. Wu, H. Yuan, L. Zhang, Y. Zhang, Recent advances on multidimensional liquid chromatography–mass spectrometry for proteomics: From qualitative to quantitative analysis—A review, *Analytica Chimica Acta*. 731 (2012) 1–10. doi:10.1016/j.aca.2012.04.010.
73. V. Spicer, P. Ezzati, H. Neustaeter, R.C. Beavis, J.A. Wilkins, O.V. Krokhin, 3D HPLC-MS with reversed-phase separation functionality in all three dimensions for large-scale bottom-up proteomics and peptide retention data collection, *Analytical Chemistry*. 88 (2016) 2847–2855. doi:10.1021/acs.analchem.5b04567.
74. Q. Yang, S. Tang, J. Rang, M. Zuo, X. Ding, Y. Sun, P. Feng, L. Xia, Detection of toxin proteins from bacillus thuringiensis strain 4.0718 by Strategy of 2D-LC–MS/MS, *Current Microbiology*. 70 (2015) 457–463. doi:10.1007/s00284-014-0747-9.
75. H. Wang, S. Sun, Y. Zhang, S. Chen, P. Liu, B. Liu, An off-line high pH reversed-phase fractionation and nano-liquid chromatography–mass spectrometry method for global proteomic profiling of cell lines, *Journal of Chromatography B*. 974 (2015) 90–95. doi:10.1016/j.jchromb.2014.10.031.
76. V.R. Richard, D. Domanski, A.J. Percy, C.H. Borchers, An online 2D-reversed-phase—Reversed-phase chromatographic method for sensitive and robust plasma protein quantitation, *Journal of Proteomics*. 168 (2017) 28–36. doi:10.1016/j.jprot.2017.07.018.
77. J. Zhu-Shimoni, C. Yu, J. Nishihara, R.M. Wong, F. Gunawan, M. Lin, D. Krawitz, P. Liu, W. Sandoval, M. Vanderlaan, Host cell protein testing by ELISAs and the use of orthogonal methods: HCP ELISAs and orthogonal methods, *Biotechnology and Bioengineering*. 111 (2014) 2367–2379. doi:10.1002/bit.25327.
78. D.G. Bracewell, R. Francis, C.M. Smales, The future of host cell protein (HCP) identification during process development and manufacturing linked to a risk-based management for their control: The future of Host Cell Protein (HCP) identification, *Biotechnology and Bioengineering*. 112 (2015) 1727–1737. doi:10.1002/bit.25628.
79. F. Yang, Y. Shen, D.G. Camp, R.D. Smith, High-pH reversed-phase chromatography with fraction concatenation for 2D proteomic analysis, *Expert Review of Proteomics*. 9 (2012) 129–34. doi:10.1586/epr.12.15.
80. R.G. Soderquist, M. Trumbo, R.A. Hart, Q. Zhang, G.C. Flynn, Development of advanced host cell protein enrichment and detection strategies to enable process relevant spike challenge studies, *Biotechnology Progress*. 31 (2015) 983–9. doi:10.1002/btpr.2114.

81. Q6B Specifications: Test Procedures and Acceptance Criteria for Biotechnological/Biological Products, (1999). https://www.fda.gov/downloads/drugs/guidancecomplianceregulatoryinformation/guidances/ucm073488.pdf (accessed December 1, 2017).
82. Z. Huang, G. Yan, M. Gao, X. Zhang, Array-based online two dimensional liquid chromatography system applied to effective depletion of high-abundance proteins in human plasma, *Analytical Chemistry.* 88 (2016) 2440–2445. doi:10.1021/acs.analchem.5b04553.
83. S. Kreimer, Y. Gao, S. Ray, M. Jin, Z. Tan, N.A. Mussa, L. Tao, Z. Li, A.R. Ivanov, B.L. Karger, Host cell protein profiling by targeted and untargeted analysis of data independent acquisition mass spectrometry data with parallel reaction monitoring verification, *Analytical Chemistry.* 89 (2017) 5294–5302. doi:10.1021/acs.analchem.6b04892.
84. A. Beck, L. Goetsch, C. Dumontet, N. Corvaïa, Strategies and challenges for the next generation of antibody–drug conjugates, *Nature Reviews Drug Discovery.* 16 (2017) 315–337. doi:10.1038/nrd.2016.268.
85. T. Botzanowski, S. Erb, O. Hernandez-Alba, A. Ehkirch, O. Colas, E. Wagner-Rousset, D. Rabuka, A. Beck, P.M. Drake, S. Cianférani, Insights from native mass spectrometry approaches for top- and middle-level characterization of site-specific antibody-drug conjugates, *MAbs.* 9 (2017) 801–811. doi:10.1080/19420862.2017.1316914.
86. A. Beck, G. Terral, F. Debaene, E. Wagner-Rousset, J. Marcoux, M.-C. Janin-Bussat, O. Colas, A.V. Dorsselaer, S. Cianférani, Cutting-edge mass spectrometry methods for the multi-level structural characterization of antibody-drug conjugates, *Expert Review of Proteomics.* 13 (2016) 157–183. doi:10.1586/14789450.2016.1132167.
87. B. Bobály, S. Fleury-Souverain, A. Beck, J.-L. Veuthey, D. Guillarme, S. Fekete, Current possibilities of liquid chromatography for the characterization of antibody-drug conjugates, *Journal of Pharmaceutical and Biomedical Analysis.* 147 (2018) 493–505. doi:10.1016/j.jpba.2017.06.022.
88. M. Excoffier, M.-C. Janin-Bussat, C. Beau-Larvor, L. Troncy, N. Corvaia, A. Beck, C. Klinguer-Hamour, A new anti-human Fc method to capture and analyze ADCs for characterization of drug distribution and the drug-to-antibody ratio in serum from pre-clinical species, *Journal of Chromatography B.* 1032 (2016) 149–154. doi:10.1016/j.jchromb.2016.05.037.
89. C. Nowak, J. K. Cheung, S. M. Dellatore, A. Katiyar, R. Bhat, J. Sun, G. Ponniah, A. Neill, B. Mason, A. Beck, H. Liu, Forced degradation of recombinant monoclonal antibodies: A practical guide, *MAbs.* 9 (2017) 1217–1230. doi:10.1080/19420862.2017.1368602.
90. J. Marcoux, T. Champion, O. Colas, E. Wagner-Rousset, N. Corvaïa, A. Van Dorsselaer, A. Beck, S. Cianférani, Native mass spectrometry and ion mobility characterization of trastuzumab emtansine, a lysine-linked antibody drug conjugate: Native MS and IM-MS for Trastuzumab Emtansine Analysis, *Protein Science.* 24 (2015) 1210–1223. doi:10.1002/pro.2666.
91. J.P. Stephan, K.R. Kozak, W.L.T. Wong, Challenges in developing bioanalytical assays for characterization of antibody–drug conjugates, *Bioanalysis.* 3 (2011) 677–700. doi:10.4155/bio.11.30.
92. R.J. Sanderson, M.A. Hering, S.F. James, M.M.C. Sun, S.O. Doronina, A.W. Siadak, P.D. Senter, A.F. Wahl, In vivo drug-linker stability of an anti-CD30 dipeptide-linked Auristatin immunoconjugate, *Clinical Cancer Research.* 11 (2005) 843–852.
93. R.V.J. Chari, K.A. Jackel, L.A. Bourret, S.M. Derr, B.M. Tadayoni, K.M. Mattocks, S.A. Shah, C. Liu, W.A. Blättler, V.S. Goldmacher, Enhancement of the selectivity and antitumor efficacy of a CC-1065 analogue through immunoconjugate formation, *Cancer Research.* 55 (1995) 4079–4084.

94. R.S. Greenfield, T. Kaneko, A. Daues, M.A. Edson, K.A. Fitzgerald, L.J. Olech, J.A. Grattan, G.L. Spitalny, G.R. Braslawsky, Evaluation in vitro of adriamycin immunoconjugates synthesized using an acid-sensitive hydrazone linker, *Cancer Research.* 50 (1990) 6600–6607.
95. M.S. Fleming, W. Zhang, J.M. Lambert, G. Amphlett, A reversed-phase high-performance liquid chromatography method for analysis of monoclonal antibody–maytansinoid immunoconjugates, *Analytical Biochemistry.* 340 (2005) 272–278. doi:10.1016/j.ab.2005.02.010.
96. R.E. Birdsall, S.M. McCarthy, M.C. Janin-Bussat, M. Perez, J.-F. Haeuw, W. Chen, A. Beck, A sensitive multidimensional method for the detection, characterization, and quantification of trace free drug species in antibody-drug conjugate samples using mass spectral detection, *MAbs.* 8 (2016) 306–317. doi:10.1080/19420862.2015.1116659.
97. T. Chen, Y. Chen, C. Stella, C.D. Medley, J.A. Gruenhagen, K. Zhang, Antibody-drug conjugate characterization by chromatographic and electrophoretic techniques, *Journal of Chromatography B.* 1032 (2016) 39–50. doi:10.1016/j.jchromb.2016.07.023.
98. J. Wang, T. Chen, C.D. Medley, K. Zhang, HPLC 2016:*Application of Comprehensive and Multiple Heart-Cutting Two-Dimensional Liquid Chromatography in Antibody-Antibiotic Conjugate Linker Drug Analysis*, San Francisco, CA, 2016.
99. S.H. Yang, J. Wang, K. Zhang, Validation of a two-dimensional liquid chromatography method for quality control testing of pharmaceutical materials, *Journal of Chromatography A.* 1492 (2017) 89–97. doi:10.1016/j.chroma.2017.02.074.
100. C.J. Venkatramani, S.R. Huang, M. Al-Sayah, I. Patel, L. Wigman, High-resolution two-dimensional liquid chromatography analysis of key linker drug intermediate used in antibody drug conjugates, *Journal of Chromatography A.* 1521 (2017) 63–72. doi:10.1016/j.chroma.2017.09.022.
101. K. Zhang, J.D. Pellett, A.S. Narang, Y.J. Wang, Y.T. Zhang, Reactive impurities in large and small molecule pharmaceutical excipients—A review, *TrAC Trends in Analytical Chemistry.* (2017). doi:10.1016/j.trac.2017.11.003.
102. L. Dai, G.K. Yeh, Y. Ran, P. Yehl, K. Zhang, Compatibility study of a parenteral microdose polyethylene glycol formulation in medical devices and identification of degradation impurity by 2D-LC/MS, *Journal of Pharmaceutical and Biomedical Analysis.* 137 (2017) 182–188. doi:10.1016/j.jpba.2017.01.036.
103. A. Tomlinson, B. Demeule, B. Lin, S. Yadav, Polysorbate 20 degradation in biopharmaceutical formulations: Quantification of free fatty acids, characterization of particulates, and insights into the degradation mechanism, *Molecular Pharmaceutics.* 12 (2015) 3805–3815. doi:10.1021/acs.molpharmaceut.5b00311.
104. D. Hewitt, M. Alvarez, K. Robinson, J. Ji, Y.J. Wang, Y.-H. Kao, T. Zhang, Mixed-mode and reversed-phase liquid chromatography–tandem mass spectrometry methodologies to study composition and base hydrolysis of polysorbate 20 and 80, *Journal of Chromatography A.* 1218 (2011) 2138–2145. doi:10.1016/j.chroma.2010.09.057.
105. O.V. Borisov, J.A. Ji, Y.J. Wang, F. Vega, V.T. Ling, Toward understanding molecular heterogeneity of polysorbates by application of liquid chromatography–mass spectrometry with computer-aided data analysis, *Analytical Chemistry.* 83 (2011) 3934–3942. doi:10.1021/ac2005789.
106. B.A. Kerwin, Polysorbates 20 and 80 used in the formulation of protein biotherapeutics: Structure and degradation pathways, *Journal of Pharmaceutical Sciences.* 97 (2008) 2924–2935. doi:10.1002/jps.21190.
107. J. Yao, D.K. Dokuru, M. Noestheden, S.S. Park, B.A. Kerwin, J. Jona, D. Ostovic, D.L. Reid, A quantitative kinetic study of polysorbate autoxidation: The role of unsaturated fatty acid ester substituents, *Pharmaceutical Research.* 26 (2009) 2303–2313. doi:10.1007/s11095-009-9946-7.

108. K. Zhang, X. Liu, Mixed-mode chromatography in pharmaceutical and biopharmaceutical applications, *Journal of Pharmaceutical and Biomedical Analysis*. 128 (2016) 73–88. doi:10.1016/j.jpba.2016.05.007.
109. A. Goyon, V. D'Atri, O. Colas, S. Fekete, A. Beck, D. Guillarme, Characterization of 30 therapeutic antibodies and related products by size exclusion chromatography: Feasibility assessment for future mass spectrometry hyphenation, *Journal of Chromatography B*. 1065–1066 (2017) 35–43. doi:10.1016/j.jchromb.2017.09.027.
110. A. Goyon, M. Excoffier, M.-C. Janin-Bussat, B. Bobaly, S. Fekete, D. Guillarme, A. Beck, Determination of isoelectric points and relative charge variants of 23 therapeutic monoclonal antibodies, *Journal of Chromatography B*. 1065–1066 (2017) 119–128. doi:10.1016/j.jchromb.2017.09.033.
111. B. Bobály, A. Beck, J. Fekete, D. Guillarme, S. Fekete, Systematic evaluation of mobile phase additives for the LC–MS characterization of therapeutic proteins, *Talanta*. 136 (2015) 60–67. doi:10.1016/j.talanta.2014.12.006.
112. A. Goyon, V. D'Atri, B. Bobaly, E. Wagner-Rousset, A. Beck, S. Fekete, D. Guillarme, Protocols for the analytical characterization of therapeutic monoclonal antibodies. I – Non-denaturing chromatographic techniques, *Journal of Chromatography B*. 1058 (2017) 73–84. doi:10.1016/j.jchromb.2017.05.010.
113. B. Bobály, V. D'Atri, A. Beck, D. Guillarme, S. Fekete, Analysis of recombinant monoclonal antibodies in hydrophilic interaction chromatography: A generic method development approach, *Journal of Pharmaceutical and Biomedical Analysis*. 145 (2017) 24–32. doi:10.1016/j.jpba.2017.06.016.
114. V. D'Atri, S. Fekete, A. Beck, M. Lauber, D. Guillarme, Hydrophilic interaction chromatography hyphenated with mass spectrometry: A powerful analytical tool for the comparison of originator and biosimilar therapeutic monoclonal antibodies at the middle-up level of analysis, *Analytical Chemistry*. 89 (2017) 2086–2092. doi:10.1021/acs.analchem.6b04726.

# 3 Solvation Processes on Reversed-Phase Stationary Phases

*Szymon Bocian*

## CONTENTS



## 3.1 INTRODUCTION

Solvation is the process of attraction and association of solvent molecules with molecules or ions of a solute. This term may also be applied for processes that take place on the interface between a liquid surface and a solid surface. In the case of liquid chromatography, during chromatographic elution, the stationary phase comes in the contact with the mobile-phase components. In such situation, molecules of the solvent are attracted by the surface. As a result of intermolecular interactions, a part of solvent molecules accumulates near the surface of the stationary phase and/or adsorbs on them. Thus, the composition of the mobile phase at the surface is different from its bulk composition. It is commonly accepted that the accumulation of molecules near the stationary-phase surface is governed by the adsorption onto the surface (adsorption model) or partitioning of the analytes between the layer of bonded ligands and the mobile phase (partition model) [1–3]. The combination of these two mechanisms is also possible [4] and most likely. Nevertheless, at present time, this mechanism is equivocal: the formation of adsorbed layer and diffuse layer, as well as partition, is considered.

The equilibration of chromatographic column after changing the mobile phase is directly connected with changes of solvation at the stationary-phase/mobile-phase interface. It is extremely important in the case of gradient elution methods. Only the analyses in stable solvation condition provide repeatable results of chromatographic analyses.

The description and discussion of solvation process are usually connected with the excess adsorption of solvents. Owing to the fact that chromatographic column is fulfilled with mobile phase, only the excessively adsorbed amount of mobile-phase component has a significant influence on the surface processes. Without the attraction of solvent molecules by the surface, the composition of the mobile phase on the stationary-phase/mobile-phase interface should be the same as the bulk mobile phase. Thus, the amount of preferentially adsorbed solvent is crucial from the solvation point of view. The discussion about the bulk mobile phase is usually omitted.

The elution strength of solvent is directly related with the ability of solvent molecules to adsorb on the stationary phase. This ability for adsorption causes stronger or weaker competitiveness between the solute molecules and, as a result, the ability for the elution of solute from the stationary phase. For these reasons, the solvation processes that take place in chromatographic column are discussed in this work.

## 3.2 SOLVATION OF STATIONARY-PHASE SURFACE

The main influence on the retention of sample components in liquid chromatography has the ability of the molecule for interaction (adsorption or partition) with the stationary phase. In isocratic chromatography, the stationary phase is in equilibrium with the mobile-phase components. This means that, in any part of the column, the composition of the components is close to thermodynamic equilibrium [5]. The equilibrium surface concentration of individual components has to be determined by corresponding/appropriate adsorption isotherms [6]. In gradient chromatography, the stationary phase is also in equilibrium with the mobile phase, but the equilibrium changes during the elution.

The components of the hydro-organic mobile phases in contact with the surface of stationary phase are preferentially distributed between the mobile and stationary phases [7,8]. It is an effect of intermolecular interaction such as Van der Waals forces, London forces, and hydrogen bond creation. The type of interaction depends on the properties of solvent molecules and the characteristic of the surface; thus, it differs between reversed-phase liquid chromatography (RPLC), normal-phase liquid chromatography (NPLC), and hydrophilic interaction liquid chromatography (HILIC) that may be considered as a form of NPLC.

### 3.2.1 Solvation in Reversed-Phase Liquid Chromatography

In RPLC with the binary (water-organic) mobile phase, the excess amount of organic modifier occurs at the stationary-phase/mobile-phase interface in a wide range of mobile-phase compositions [9]. The quantity of excess amount of the adsorbed organic modifier depends on the concentration of the modifier in the eluent, and it is the lowest at a high concentration of organic solvent in the mobile phase. The total amount of adsorbed organic solvent depends on many factors, for example, the length and the number of organic ligands [10], the coverage density [11], and the presence of polar group in the ligand structure [12]. The influence of particular parameters on the solvation processes will be described in detail in the further part of this chapter. Description of solvation processes is more complicated in the case of ternary mobile phases.

In binary mobile phase, when the concentration of the organic modifier in the mobile phase decreases, its excessively adsorbed amount is higher, which results from the mobile-phase composition. In this case, an excess amount of solvent is observed for lower concentration of organic solvent in the mobile phase. However, total concentration of a given solvent on the stationary-phase surface increases with its increase in the mobile phase, except for extreme concentrations, usually higher than 80%. In RPLC, the excess amount of adsorbed water may also be observed. In order to investigate this effect, the concentration of water must be usually lower than 20%. In this condition, water strongly adsorbs on residual silanols. This adsorption is so strong that water molecules may form a multilayer structure [7]. In the excess isotherm measurement of organic solvent, the water adsorption is observed as a negative part of the excess isotherm. Water molecules create a structure called "hydrophilic pillow" [13–16]. Hydrophilic pillow is observed in the case of a significant amount of unshielded residual silanols or when polar groups are incorporated in the structure of chemically bonded stationary phase. As an example, acetonitrile (ACN) excess isotherm on alkyl-amide stationary phase is presented in Figure 3.1. Negative part of the isotherm corresponds to the adsorption of water that causes the creation of hydrophilic pillow near the silica support surface.

Another assumption made in the discussion of solvation processes is that the stationary surface is flat. It allows the calculation of the theoretical number of adsorbed monolayers of solvent. It has to be underlined that such calculations have physical meaning only on unmodified surfaces, for example, unmodified silica gel. In the case of chemically bonded stationary phase, such description is only an assumption [17,4].

In the case of ACN as an organic modifier in RPLC—which exhibits stronger adsorption than methanol—the hydrophobic effect plays the most significant role in its adsorption [4]. ACN preferentially solvates organic ligands (alkyl or aryl). Owing to the inability of ACN to create hydrogen bonds, ACN does not solvate residual silanols. It results in the preferential adsorption of water. In the case of methanol,

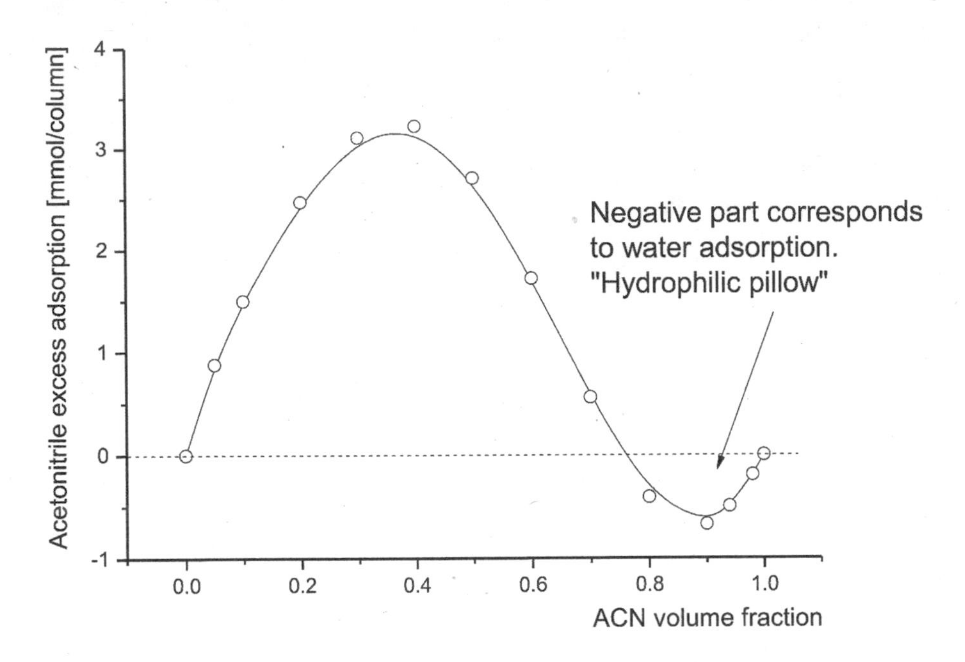


**FIGURE 3.1** Excess adsorption of acetonitrile on alkyl-amide stationary phase. An evidence of "hydrophilic pillow."

the situation is more complicated owing to the possibility of interaction of methanol molecules by forming hydrogen bonds [18].

On reversed-phase stationary phases, the adsorption of ACN is stronger than the adsorption of methanol. However, water solvates residual silanols stronger from ACN-water mixtures than from the mixtures with methanol. As a result, negative parts of the excess isotherm are more essential for ACN-water mixtures than for MeOH-water mixtures. The solvation regions of the RP stationary phases are presented in Figure 3.2.

Similar situation is observed in the case of two less popular solvents in RPLC: 2-propanol and tetrahydrofuran (THF). Comparing two alcohols, the amount of adsorbed 2-propanol is about 1.5 times larger than that for methanol on the C18 stationary phases. It is a result of higher hydrophobicity of 2-propanol. For water-rich mobile phase, the excess amount of adsorbed alcohols increases with the increasing number of carbon atoms in alcohol molecules, from methanol to ethanol and 2-propanol, as it is expected in reversed-phase system. For mobile phases rich in organic modifier, the excess adsorption of water increases from methanol solution to 2-propanol-water mixture. The more hydrophobic and the larger alcohol molecules cannot penetrate into the bonded ligands, and their competition with water for adsorption onto the silanol groups is weaker [17].

For the water-rich mobile phase, the amount of adsorbed THF is usually higher than that for ACN. Adsorption of alcohols is lower than that of non-alcoholic organic modifiers such as ACN and THF [17]. Exemplary excess isotherms of various solvents on the C18 stationary phase are presented in Figure 3.3.

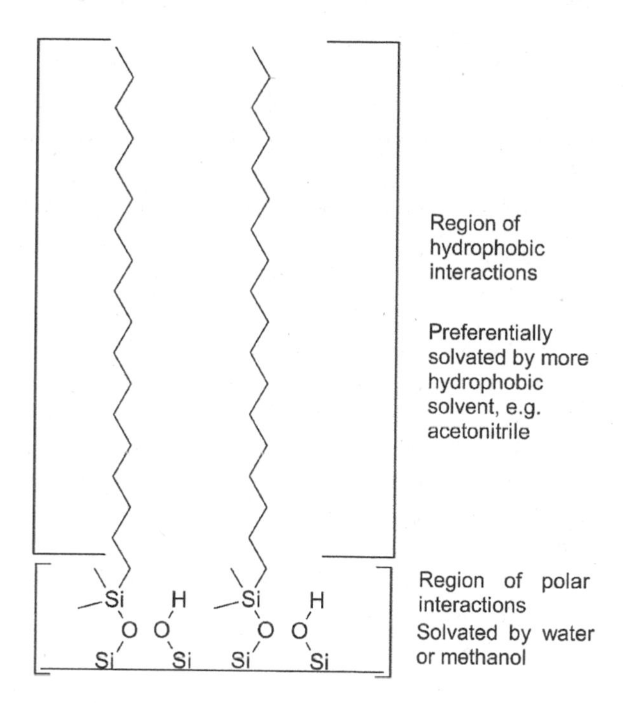


**FIGURE 3.2** Solvation region of RP stationary phase.

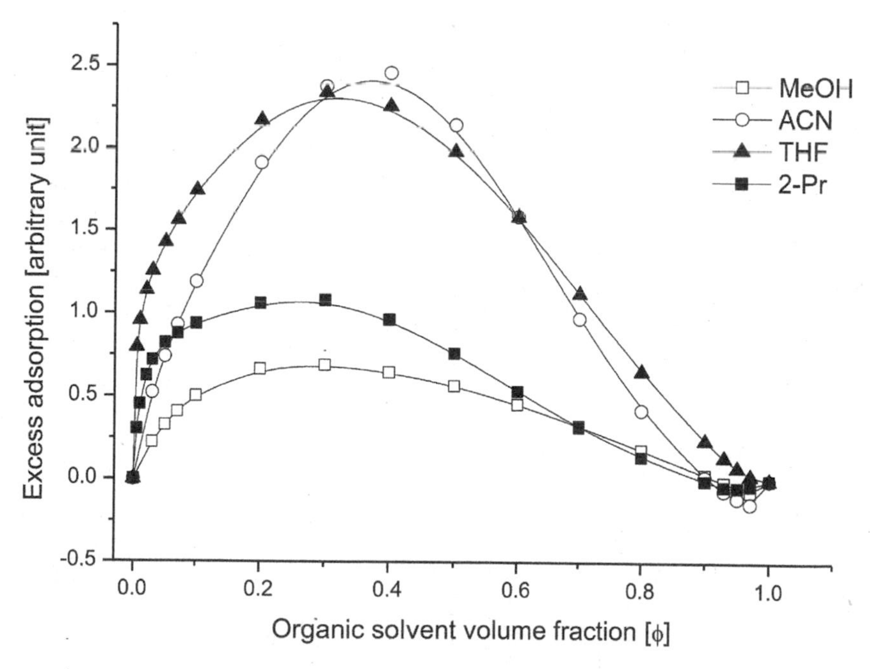


**FIGURE 3.3** Solvent excess adsorption isotherms measured on C18 stationary phase (relative values).

Similar solvation processes to high-performance liquid chromatography (HPLC) take place in supercritical fluid chromatography (SFC). In SFC, the excess of adsorbed organic modifier is observed, and it may be measured using analogous method. Modifier adsorption is determined from carbon dioxide solution [19–22].

### 3.2.2 Solvation in Hydrophilic Interaction Liquid Chromatography

In HILIC, opposite situation is observed. Stationary-phase surface is polar. On the polar surface of the stationary phase, the adsorption of more polar component takes place; therefore, from ACN-water mobile phase, water molecules are preferentially adsorbed [15,16,23,24]. In the literature, the discussion of solvation processes in HILIC is limited to ACN-water solution; however, other organic solvents were tested for HILIC separations.

Preferential adsorption of water molecules in HILIC on bare silica gel was also investigated by Guiochon et al. [25] and Vajda et al. [23]. In all of the above-mentioned measurements, an excess of adsorbed water was observed at silica surface. The adsorption of water on silica surface was also confirmed by computer modeling by Melnikov et al. [26]. It is obvious that similar situation should take place when other polar stationary phases are used in HILIC. Preferential adsorption of water near the stationary phase's surface causes the creation of the so-called hydrophilic pillow that plays a crucial role in HILIC retention mechanism.

As a result of the preferential solvation of the stationary-phase surface, it is very likely that the gradient of a given solvent concentration from the adsorbent surface into a bulk mobile phase is formed [16]. Therefore, the composition of the mobile phase at the interface is different from its bulk composition. From the retention mechanism point of view, it is still difficult to recognize whether the mechanism of solute distribution is adsorption or partition [8,10]. Probably, the retention of solutes may be due to the combination of adsorption and liquid-liquid partition mechanism, which, together with possible additional effects, controls the retention [27].

It is possible that the diffuse layer of the adsorbed water constitutes a zone of polar interaction, opposite to ACN-rich mobile phase, where hydrophobic interactions are dominant. It is illustrated in Figure 3.4. Polar analytes are distributed

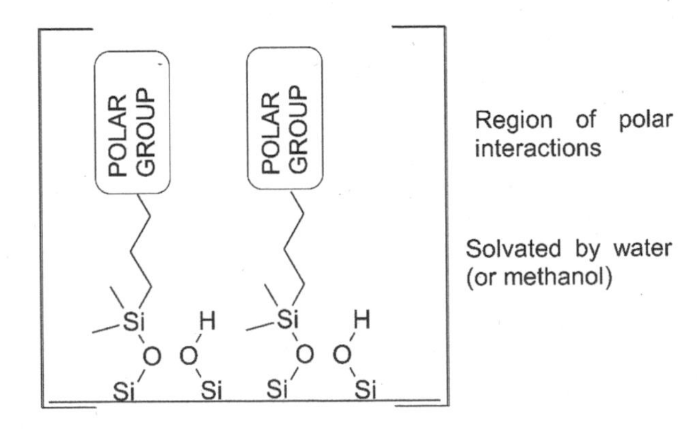


**FIGURE 3.4** Solvation region of HILIC stationary phase.

between ACN-rich mobile phase and water-rich adsorbed layer. Thus, higher amount of adsorbed water creates bigger zone of polar interaction where analytes are accumulated. It was also observed that the retention of analytes is in proportion with the amount of water adsorbed near the surface of the stationary phase [15].

Similar situation may be observed when the hydrophobic stationary phase possesses incorporated polar groups. Such groups may accumulate significant amount of water and create a region of hydrophilic interactions, whereas hydrophobic ligands constitute a region of hydrophobic interactions. In this case, such stationary phases may be applied in both RP and HILIC system, depending on the separated solutes and mobile-phase compositions [28,29]. Similar surface phenomena are also observed in normal-phase chromatography [30].

## 3.3 SOLVATION IN GRADIENT ELUTION

The description of solvation processes in isocratic elution is relatively easy to perform, because column is in equilibrium and solvation effects do not change significantly during the elution process. Only local changes are observed, caused by the adsorption on the solute molecules. More complicated situation takes place during gradient elution. In gradient elution, the concentration of stronger solvent increases during the analysis. This causes that the solvation processes are not in the equilibrium all the time. In addition, the equilibrium takes place at very short distance in the column, and at each time, it varies with the column length. To simplify, the column is divided for a series of local equilibria.

During the gradient elution, the amount of weaker solvent on the stationary phase decreases, because those molecules are replaced by the molecules of stronger solvent. In gradient elution, the solvation process is almost impossible to measure, but the most important is to perform analyses in reproducible conditions. This statement is confirmed by the lack of literature that describes the solvation phenomena in gradient elution. If the gradient elution starts on the equilibrated column, solvation changes caused by an increased concentration of stronger solvent give reproducible effect on the separation. Thus, the stable initial condition guarantees the reproducible solvation changes during gradient elution [31,32].

## 3.4 MEASUREMENT OF SOLVATION PROCESSES

The role of the mobile-phase components and their influence on the solute retention are well known [33]. However, the underlying phenomena often have not been well understood. To find a relationship between solute retention and mobile-phase composition, many attempts have been made [34–37]. One of the ways to find the thermodynamic parameters of the molecular interaction between the solvent and the stationary phase is solvent adsorption determination. In such case, the adsorption isotherm can be used. Result of solvent adsorption should give useful information about the retention of organic compounds in chromatographic columns [6]. There are various chromatographic methods (static and dynamic) for the measurement of the adsorption isotherms of mobile-phase components [6,38–43].

The accumulation of solvent molecules near the surface—so-called preferential solvation effect—for any system can be described by Gibbs isotherm:

$$\Gamma_A = -\frac{c_A}{RT} \cdot \frac{d\delta}{dc_A} \quad (3.1)$$

Where $\Gamma_A$ is the adsorbed amount of substance $A$, $c_A$ its concentration in solution, and $\delta$ the surface tension between the surface of the stationary phase and solution.

The amount of adsorbed solvent is usually determined using frontal analysis and excess isotherm measurement. It has to be emphasized that, sometimes, results from both methods do not correlate. The amount of adsorbed solvent from the gaseous phase on chromatographic adsorbent was also measured [9].

### 3.4.1 Frontal Analysis

To measure the adsorption from binary liquid mixtures, the frontal analysis may be applied. The complete adsorption isotherm of a solvent mixture is obtained by the measurement of the breakthrough curves for a series of small concentration steps of the mobile phase [44]. This method offers a direct way to determine the composition of the stationary phase in liquid-solid chromatography with mixed mobile phases. The surface excess isotherms of all binary systems formed by benzene, cyclohexane, and 1,2-dichloroethane, at the solution-silica gel interface, were measured by Köster and Findenegg [45]. Last attempt was done by Soukup and Jandera [46] to determine the adsorption of water from ACN on 16 different stationary phases. Frontal analysis method with coulometric Karl Fischer determination of water in the column effluent was performed. It has been observed that at full column saturation, the excess adsorbed water fills 2.3%–45.3% pore volume, which corresponds approximately to the equivalent of 0.25–9 water layer coverage of the adsorbent surface. It is assumed that the equivalent number of the adsorbed water layers decreases with the decreasing number of active silanols and hydroxyl groups [46]. Analogues methodology was used to determine water adsorption on hydrosilated silica-based stationary phases [47].

Although frontal analysis is more commonly used for solute adsorption isotherm determination, their application for solvent adsorption measurement is limited, because it is difficult to measure solvent adsorption in its whole concentration in the mobile phase (0%–100%). However, frontal analysis of the organic solvents provides information complementary to the excess isotherm measurements [48]. Nevertheless, excess adsorption measurement is more popular method for the determination of solvent adsorption.

### 3.4.2 Excess Isotherm

The excess adsorption isotherm of mobile-phase component (e.g., methanol, ACN, and water) from its solutions may be measured experimentally as a difference between the amounts of the component that would be in a hypothetical system without surface influence and the amounts of the component in the system with surface influence [5]. Measurements are carried out from binary solution. The measurement may be carried out if the given component adsorbs more strongly or similar to the second one. It is impossible to measure correctly the adsorption of weakly adsorbed solvent from its solution in strongly adsorbed solvent. The application of the minor disturbance method is limited to the determination of excess adsorption isotherms of electrolyte-free binary mixtures [49].

The chromatographic process based on the adsorption theory was described by Riedo and Kováts [3]. The retention volume of a component passing through the column is a result of the mass flow rate and mass transfer between the mobile- and the stationary-phase surfaces [3,50]. In this case, the transport equation allows the combination of the retention with adsorption. For two components' mixture, the universal mass balance equation may be solved analytically. This approach is based on the consideration of the total amount of the mobile-phase components in the cross-sectional area of the chromatographic column. During adsorption, each component is distributed between the bulk mobile phase and the adsorbent surface. It leads to the determination of the adsorption layer [3,5,7].

In liquid systems, the adsorption process is characterized by the excess adsorption isotherms. Wang et al. incorporated the Gibbs adsorption definition with the material balance in the dynamic chromatographic processes [51]. For a solution-solid surface, the Gibbs adsorption definition for the absolute surface excess concentration may be reduced to the following equation:

$$\Gamma_i^s = \int_0^\infty [c_i(z) - c_i^\infty] dz \tag{3.2}$$

where $c_i(z)$ is the concentration along the perpendicular from the surface, and $c_i^\infty$ is the concentration far away from the surface.

A more practical definition of the excess isotherm was given by Kazakevich and McNair [52,53] as a difference between the initial and equilibrium concentration, multiplied by the number of moles of the solute and related to the unit of surface area.

The excess isotherm can be measured using a perturbation method (in the literature named as minor disturbance method or step-and-pulse method) [3,5,6,9,54]. This method consists of introducing and measuring a small perturbation in a biphasic system under equilibrium. The concentration of organic modifier in the mobile phase is changed stepwise. The small perturbation is caused by the injection of pure modifier on each conditions, and the retention time is measured [6,8,55]. Observed perturbation peak is also called "solvent peak" [56,57].

The determination of the excess isotherms starts with measuring the thermodynamic void volume of the column by the minor-disturbance method. The equilibrium between the binary mobile phase and the adsorbed phase is perturbed, and the elution volume of the perturbation is recorded at every mobile-phase composition, from pure water to pure organic solvent. The thermodynamic void volume of the column is obtained by integrating the retention times of the perturbation peaks over the mobile-phase composition [5]:

$$V_M = \frac{1}{C_{\max}} \int_0^{C_{\max}} V_R(C)dC \tag{3.3}$$

where $V_M$ is the thermodynamic void volume of the column, $V_R$ is a retention volume of the perturbation peak, and $C$ is the concentration of the analyte (mol/L).

The dead volume ($V_m$) may be shown on the plot of the retention volume of solvent, as is demonstrated in Figure 3.1.

This volume can be used to calculate the excess adsorption isotherm of the organic modifier in the column. When $V_m$ is known, the excess amount ($\Gamma$) can be calculated (adsorbed per unit surface area of the adsorbent). Experimental excess isotherm can be calculated from the equation [5,58]:

$$\Gamma(C) = \frac{1}{S}\int_0^{C} \left(V_R(C) - V_M\right)dC \tag{3.4}$$

where $S$ is a total surface area ($m^2$).

Traditionally, excess amount of adsorbed solvent is calculated per surface unit of the stationary phase in the column. In practice, it is difficult owing to the problem with exact determination of the surface area of the packed column.

The main problem in the comparison of solvent adsorption is the accurate definition of the stationary-phase surface [5,59–61]. The amount of adsorbed solvent was calculated in different ways: calculations of surface area in the column [17], per gram of stationary phase [7,9,62], per surface of the silica support [12], per volume of the stationary phase [11,63,64], and per surface of the stationary phase after modification [4,10,58] or per column [15]. Although the first methodology is the most proper, the error obtained during the estimations of the surface area in the column is significant. Different methods of surface excess calculation make the comparison of data difficult. In addition, the precise determination of the real surface is necessary for correct interpretation of surface phenomena.

During the calculation of the surface area of stationary phase using Brunauer–Emmett–Teller (BET) theory, it has to be emphasized that for hydrophilic surfaces, 16.2 $Å^2$ surface of a nitrogen molecule is calculated for the low-temperature nitrogen adsorption (LTNA) measurement but that would underestimate the surface area values of hydrophobic surfaces. Kazakevich et al. [65] and Bass et al. [66] suggest the application of a modified surface area of a nitrogen molecule (20.5 $Å^2$) for hydrophobic surfaces.

The mass of the packing material present in chromatographic columns may be calculated using the methods presented by Gritti et al. [4] and Kazakevich et al. [65]. Then, the total surface areas of silica gel in the column may be calculated using known carbon load and specific surface area of the bare silica gel.

Organic ligands bonded to the silica support occupy some volume inside the pore space, thus decreasing the original support pore volume. In this case, the pore volume, pore diameter, and specific surface area decrease as well [59,61,67–71]. Based on the LTNA and inverse size-exclusion chromatography (ISEC) measurement, it is possible to calculate the theoretical surface area of the stationary phase covered with bonded ligand. Detailed description of this methodology is given elsewhere [17].

The measurement of the excess adsorption isotherm of a component in the binary system does not require a priori introduction of any model. It is possible to consider the excess adsorption isotherm as being model-independent, and it is possible to derive the properties of the adsorbed layer on the basis of consideration of a concentration-dependent adsorption process [53].

The amount of preferentially adsorbed component ($n_{ads}$) per unit of surface area can be calculated from the equation:

$$n_{ads} = C\tau + \Gamma(C) \tag{3.5}$$

where $\tau$ is the thickness of the adsorbed layer, which can be calculated from the following equation:

$$V_{ads} = SC_{ads}\upsilon = S\tau \tag{3.6}$$

$C_{ads}$ can be found by extrapolating the slope of excess isotherm in a linear region to the intercept y-axis, and $\upsilon$ is a molar volume of component. Detailed information about the calculation of solvent excess isotherm and the origin of the equation presented above may be found in the book of Kazakevich and LoBrutto [53].

In the calculation and interpretation of the excess isotherms, the value of surface area of the stationary phases has a significant influence on the results. The excess adsorbed amounts of solvent calculated per surface of the silica gel in the columns exhibit significant influence of the decrease of the stationary bonded-phase surface that is caused by the increase of the coverage density of bonded ligands. However, calculations performed with modeled bonded ligand surface eliminate this problem. It seems that using the modeled surface area of bonded ligands gives a more correct explanation of the solvation processes [17]. Another assumption is to calculate the solvent excess per number of bonded ligand. The number of adsorbed solvent molecules per one ligand decreases with the coverage density of bonded ligands. The thick structure of the bonded ligands makes the penetration of solvent and solute molecules from bulk mobile phase to silica gel surface difficult [17,53].

Determination of excess isotherms allows the calculation of the number of theoretical adsorbed layers of solvents. The number of adsorption monolayer can be calculated from the equation proposed by Gritti and Guiochon [4,72].

$$t = -\left(\left[\frac{dn_1^e}{dx_1^e}\right]_I \left(x_1^l a_1^* + \left[1 - x_1^l\right] a_2^*\right) + \left(a_2^* - a_1^*\right)\left[n_1^e\right]_I\right) \tag{3.7}$$

where $I$ is the inflection point of the excess adsorption isotherm observed in its decreasing branch, $x_1^l$, is the molar composition, $[n_1^e]$ is the excess adsorbed amount, and $[dn_1^e / dx_1^e]$ is the derivative of the excess adsorbed amount with respect to the molar fraction of component 1 at the inflection point $I$. Parameters $a_1$ and $a_2$ are the surface requirements per molecule of organic modifier and water, respectively, when they are adsorbed on hydrophobic surfaces. The adsorption of methanol on the chemically bonded phases is monolayer, and the adsorption of other solvents (e.g., ACN and THF) is multilayer. These results are consistent in the papers of various authors: Gritti and Guiochon [4,72], Rustamov et al. [59], and Bocian et al. [17]. Exemplary results of calculated adsorbed layers on a series of C18 stationary phases are presented in Figure 3.5.

In the case of adsorption on a homogeneous surface, the inflection point of decreasing branch of the isotherm can be observed at $x^l_1 = 1$ (type I excess isotherm without negative part—only one component from binary solution is adsorbed) [72–74]. In the case of solvent, type V and type IV excess isotherms (with significant negative part of the isotherm corresponding to adsorption of second component from binary mixture) are observed, with inflection point ranging between 0.43 and 0.58, 0.33 and 0.37, 0.18 and 0.20, and 0.23 and 0.30 for methanol, ACN, 2-propanol, and THF, respectively. This clearly demonstrates that the surfaces of the stationary phases used for liquid chromatography become heterogeneous.

The theoretical model of excess isotherms assumes a surface covered with patches of two different kinds of adsorption centers—organic ligands and residual silanols [4]. The two types of surface patches interact with the organic modifier and water in a different way while they exhibit different adsorption energies. In described model, parameters $K_{ligand}$ and $K_{silanol}$ represent the two types of adsorption sites, organic ligands and residual silanols, and parameter ε represents the surface heterogeneity.

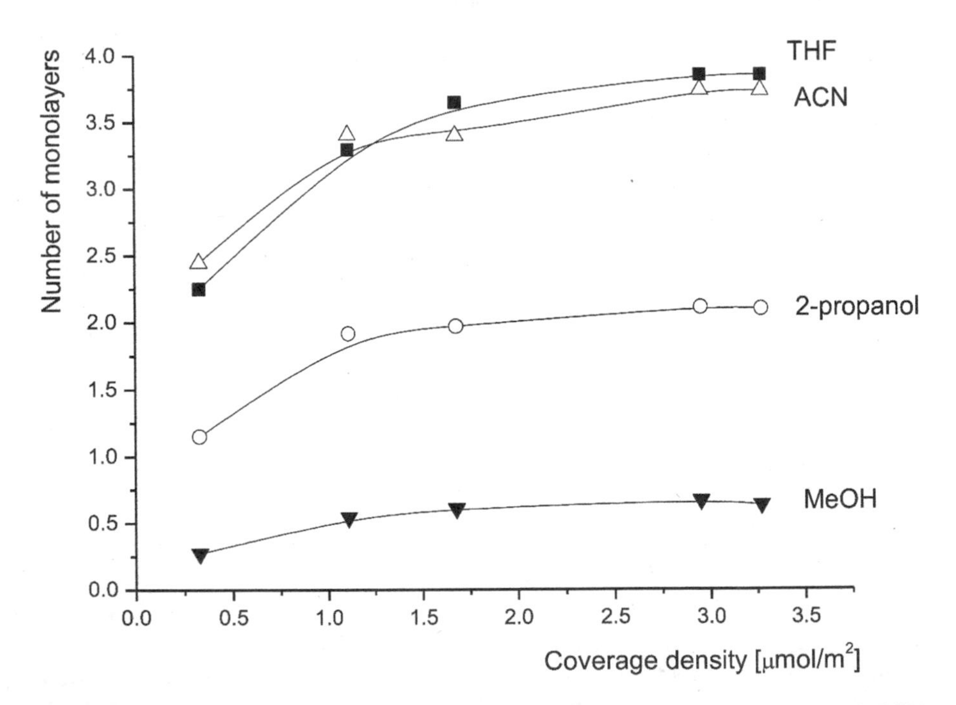


**FIGURE 3.5** Exemplary results of calculated adsorbed layers for MeOH, 2-propanol, ACN, and THF on a series of C18 stationary phases.

The following equation for the case of heterogeneous monolayer adsorption was derived by Everett [37]:

$$n_1^e = At\left(\varepsilon\frac{(K_{ligand}-1)x_1^l(1-x_1^l)}{K_{ligand}a_1^*x_1^l + a_2^*(1-x_1^l)} + [1-\varepsilon]\frac{(K_{silanol}-1)x_1^l(1-x_1^l)}{K_{silanol}a_1^*x_1^l + a_2^*(1-x_1^l)}\right) \quad (3.8)$$

Because of the simplifying assumptions made, and since the adsorption of most of the organic solvents is multilayer, the fit of Equation 3.8 can only give qualitative results [4,72].

Parameter $K_{ligand}$ increases with increasing coverage density of organic ligands. This is because of the increase of the adsorption potential when increasing the density of the C18 chains bonded to the surface. However, the surface heterogeneity of the stationary phase does not reduce significantly with the increase of the degree of silica surface coverage with bonded ligands [17].

### 3.4.3 Microcalorimetric Study in Solvation Processes Investigations

The microcalorimetric measurement of the thermal effect accompanying wetting of the stationary phase by organic solvent is another method that can give useful information about solvent interactions with the stationary bonded-phase surface. The microcalorimetry measurement can help to define the conformation of chemically bonded phases and to predict the possibility of solvent penetration through the film of bonded ligands. This penetration between ligands, and thus solvation process, depends strongly on the organization of bonded ligands on the surface of the support [75–77].

The measured heat of adsorption is generated by both specific and non-specific interactions. All of the tested solvents, methanol, ACN, and hexane, may act with the silica surface owing to non-specific interactions (van der Waals forces and London forces), but only methanol and ACN exhibit specific polar interactions (hydrogen bonding and dipole-dipole interactions). As a result, the heat of adsorption is higher for alcohols than for hydrocarbons [78,79].

Changes of the heat of immersion inform about the accessibility of the surface for interaction with solvent molecules. The increase of the stationary-phase coverage density reduces free space between bonded chains. The following regularity was observed for hydrophobic stationary phases: an increase in the coverage density of the stationary phase causes a decrease of the number of adsorbed molecules per one organic chain for methanol, ACN, and hexane. This suggests that the penetration of solvent molecules through a dense film of bonded ligands is more difficult. The residual silanols are more effectively screened by long chains of the stationary phase [80]. Microcalorimetric measurements may be used for the determination of the polarity of stationary phases and their hydrophobic-hydrophilic properties [81]. Measurements of immersion heat, especially for methanol, which can create hydrogen bonding with residual silanols, provide useful information about the accessibility of the silanols for interaction with the solute during chromatographic elution [82]. Calorimetric measurement may be applied for the determination of the number of accessible silanols. Based on the difference in thermal effects of the surface immersion of methanol and hexane, the number of hydrogen bonds created can be calculated. It can be also used for bonded stationary-phase characterization by determining the number of silanols that are available for hydrogen bond interactions during chromatographic elution [83].

### 3.4.4 Inverse Method of Isotherm Determination

If the mass-transfer kinetics is fast and the probe molecules have small molecular weight, the mass balance of the chromatographic system can be modeled with the equilibrium-dispersive model [41]. The boundary conditions of this model are modeled with a exponentially modified gaussian function (EMG) to account for the proper injection profile.

The integration of the mass-balance equation is carried out using the finite-difference Rouchon method [41] to calculate band profiles and compare the experimental and theoretical overloaded bands of the solvent molecules. The parameters derived from the frontal analysis data may be used to model overloaded elution bands without any optimization or fitting procedure [48]. This method is not commonly used for the determination of solvent adsorption.

### 3.4.5 Alternative Methods of Solvation Process Determination

An alternative method for determining excess adsorption isotherms is the tracer pulse method, using a mass spectrometer [19,84]. A tracer pulse peak is generated by an isotopically labeled eluent component. This peak is detected by the mass spectrometer. When the minor disturbance method and the tracer pulse method are compared, the resulting excess adsorption isotherms are similar [85].

Another method for the reliable measurement of excess adsorption isotherms of organic eluent components were proposed by Ohashi et al. [49]. The method uses an HPLC system with photodiode array detector (PDA) for determining adsorption isotherms. The excess adsorption isotherms of organic solvents are determined using the elution volumes of peaks produced by the refractive index change at the breakthrough curve in frontal analysis. This method was applied to the measurement of the excess adsorption isotherms from the mobile phase containing electrolyte ammonium acetate [49]. This method eliminates the problem of peaks generated on the chromatogram that are connected with additives, such as salts, during excess isotherm measurement from mobile phases containing additives.

Adsorption energy distribution (AED) allows to choose the proper model that describes the adsorption behavior of the studied compound and gives an alternative way to determine the isotherm model parameters from the raw adsorption isotherm data. The advantage of the method is the ability to estimate the heterogeneity of the stationary phase, and the isotherm parameters are derived from the raw isotherm data, without any further assumption [48]. The detailed description of the numerical procedure (expectation maximization algorithm) and methodology can be found in the literature [86].

The calculation of AEDs confirmed that, for protic solvents, bimodal distributions are observed only at the highest surface coverage of octadecyl chains. At low surface coverage, the AED is unimodal. On the contrary to protic solvents, the adsorption of THF is always heterogeneous on the octadecyl phases. With the decrease of surface coverage, the high-energy sites become more abundant—in parallel with the increase of the gap between the bonded chains. Similar tendencies are observed for the low-energy sites, and that is related to the bigger pore volume, and surface is available for adsorption. The AED is unimodal on the homogeneous surfaces, e.g., C1 modified adsorbent [48].

### 3.4.6 Computer Modeling of Solvation Processes

Computer simulations provide a new level of insight into the retention process in liquid chromatography. Data obtained from the computer modeling may help in a complete or better understanding of the elution process and improve the knowledge in the field of analytical separation sciences [87,88]. Computer modeling is a powerful tool for the description of the solvation processes that take place in RPLC. For example, the particle-based Monte Carlo simulations were employed to examine the effects of the bonding density of the molecular structure in RPLC [89]. The modeling of ligands and the conformation and partitioning of solvent in a RPLC system were also studied [90–92].

In the case of the comparison of computer modeling and measurement, the results from the computer optimizations using CACHE Suite program correlate well with the data obtained from the measurement of solvent adsorption. For the lowest-coverage-density column, the amount of the adsorbed ACN was ideally consistent with the computer simulation. For the stationary bonded phases with higher surface coverage, larger differences were observed, because the modeled surface cannot represent exactly the structure of the bonded phase [91]. An example of modeled solvation process on C18 ligand is presented in Figure 3.6. The modeling of solvation processes was also done for HILIC systems [93,94].

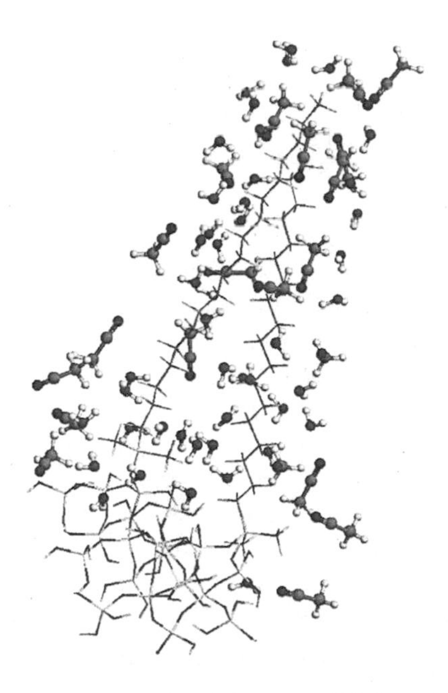

**FIGURE 3.6** Visualization of computer simulation: two C18 chains bonded with silica surface in ACN/water environment. (Based on Buszewski, B. et al., *J. Sep. Sci.*, 33, 2060–2068, 2010.)

## 3.5 PARAMETERS INFLUENCING SOLVATION

The amount of adsorbed solvent depends on the type of the solvent. In RP systems, it is assumed that the adsorption of methanol on the hydrophobic stationary phase is usually monolayer [4]. ACN can adsorb on the stationary phase, forming three or more layers [4,5]. These results are consistent with solvent adsorption data on stationary bonded phases measured by gas chromatography [9]. In addition, solvent adsorption in RP HPLC system strongly depends on the temperature. A significant difference is observed in the shape of the excess isotherm. The amount of adsorbed solvent on the stationary phase decreases with the increase of the temperature [5,95,96].

### 3.5.1 Coverage Density of Bonded Ligands

The adsorption of organic solvent depends on the coverage density of bonded ligands [4,11,59,97]. Overall, higher surface coverage with hydrophobic ligands causes higher hydrophobicity of the surface. More hydrophobic surface alters the adsorption of organic solvents. However, in detailed description of the surface processes, the changes of the surface inside the column have to be taken into account. The chemical modification of the silica gel by binding organic ligands changes the geometry of the silica support and influences the solvation processes and the separation mechanism [59,67,98]. After derivatization of the silica support, bonded organic ligands occupy some volume inside the pores. This effect decreases the original pore volume of the silica gel. If the pores are filed with bonded ligands, the specific surface area decreases as well [61,65,67–69,99]. The degree of the volume and the surface area reduction inside the pores depend on the total volume of bonded ligands (the length of the organic ligands and their coverage density). Rustamov et al. [59] determined that the pore volume decreases by 18 μL/g per $CH_2$ group and the volume of a C18 ligand as about 600 Å$^3$. The surface area of bonded phase after derivatization may be estimated based on the carbon load, porosity of the material, density parameters, etc. [4,10,58].

In the series of the columns with various surface coverage, as the surface coverage increases, the excess amount of adsorbed solvent also increases, until the effect of the decreasing surface area becomes significant. However, using the calculated surface area of the stationary phase instead of silica gel surface, the solvent excess amount increases in whole range of coverage density, as in Figure 3.7. It suggests that organic modifier molecules cannot penetrate the bonded ligands if their coverage density is high. They adsorb only on the top of the ligand [17]. In addition, bonded ligands in the stationary phase may change the conformation when mobile phase is changed, for example, hydrophobic ligand may collapse when the concentration of water in the mobile phase is high. It also has to be emphasized that the bonded phase does not participate in retention as a homogenous entity but contains multiple sorption centers and changes the composition of the interfacial region when it is in contact with various mobile phases [88].

### 3.5.2 Type and Length of Bonded Ligand

The excess amount of the adsorbed organic solvent on the stationary phase depends on the type of bonded ligands [5,7,11,63,100] as well as the length of the organic

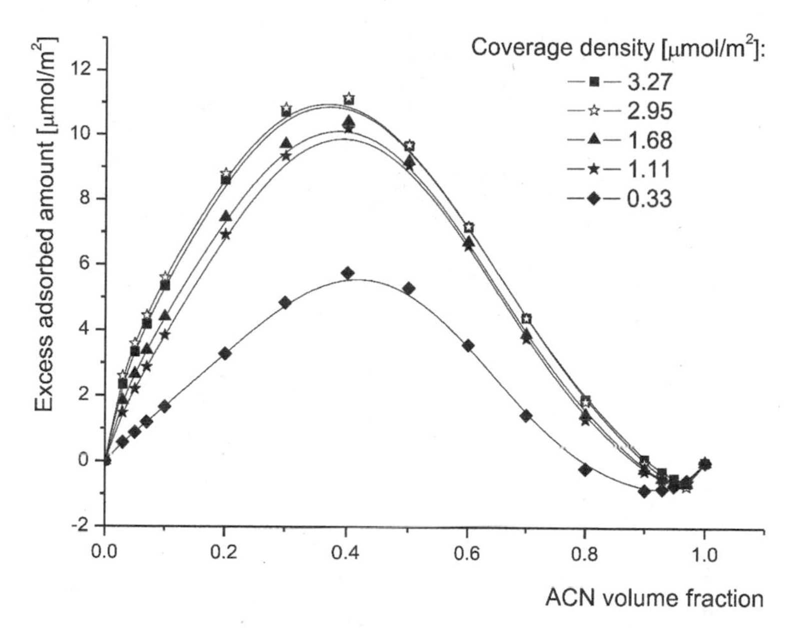


**FIGURE 3.7** Effect of C18 ligand coverage density on acetonitrile adsorption. (Based on Bocian, S. et al., *Anal. Chem.* 81:6334–6346, 2009.)

chains of the stationary phase [10,58,59]. The presence of different types of functional group in the ligands' structure has a significant influence on the solvent adsorption. Maximum adsorbed solvent increases with the number of carbon atoms from the amino propyl ligands to alkylamide and cholesteryl stationary phase [12]. In comparison to methanol, ACN is preferentially adsorbed on the ligands that exhibit the ability for π-π interaction such as phenyl bonded material and the stationary phase with cyano groups [12]. On the other hand, methanol is preferentially adsorbed on the alkylamide and amino stationary phases *via* hydrogen bonding and other proton-donor and proton-acceptor interactions. The solvent adsorption governed by these specific interactions may have substantial influence on the separation selectivity in the discussed systems. Exemplary excess isotherms of ACN on various stationary phases are presented in Figure 3.8.

The presence of polar functional group causes the relative elution strength of organic solvent to change, depending on the stationary phase. Polar functional groups, such as amine, hydroxyl (diol), amide, and ester groups, constitute a center of polar interaction, which increases the stability of the stationary phase in water-rich mobile phases in RP system as well as allows to apply such stationary phase in HILIC [28].

The presence of polar functional groups changes the relative elution strength of methanol and ACN. In the case of iso-eluotropic mobile phases, ACN-water mixtures exhibit relatively higher elution strength on phenyl-bonded stationary phase than MeOH-water mobile phases. However, if the phenyl-bonded stationary phase

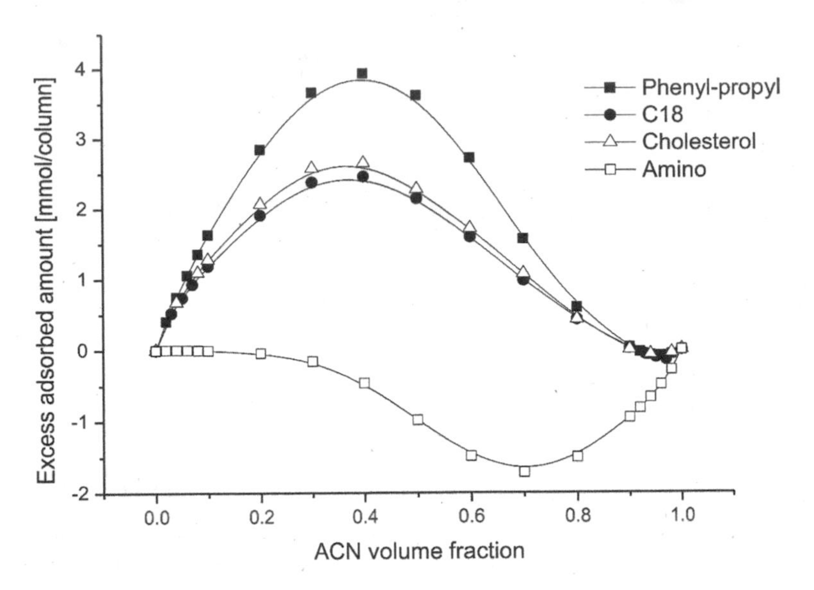


**FIGURE 3.8** Exemplary excess isotherms of ACN on various stationary phases.

possesses amine or amide groups in the ligand structure, the application of MeOH-water mobile phase causes shorter retention than ACN-water [101].

### 3.5.3 End-Capping

Despite the derivatization process of the silica gel during stationary-phase synthesis, some portion of the residual silanols is always present. The presence of accessible residual silanols can have a negative influence on the separation of polar analytes, especially basic compounds and biopolymers [102,103], because of polar interactions. From the surface properties' point of view, the stationary-phase surface becomes heterogeneous. The heterogeneity of the bonded phase decreases with the increase of the coverage density of bonded ligands and depends on the organic solvent in the mobile phase [4]. For the bonded stationary phases with high coverage density of bonded ligands, from the amount of residual silanols—which are not bonded with silanes—only about 5% exist as a polar strong-adsorption center—accessible residual silanols [104]. This effect is caused by shield properties of the methyl (or other, e.g., isopropyl) group connected to the silicon atom in the silanes, used for silica gel modification [104–106]. Weak homogeneity and low coverage density of the bonded ligands can uncover the higher surface of the silica gel and expose more accessible silanols for polar interaction, which worsen chromatographic separation and change solvation processes owing to the increasing number of polar adsorption sites [76].

Usually, excess isotherms of ACN in RP system have a small negative part for high concentration of ACN in the mobile phase. This negative part is caused by the adsorption of water on the residual silanols [4,5,10,11,40,53]. This part of the isotherm is much bigger, when the ACN excess isotherm is measured in HILC owing to the adsorption of water on various polar group in the stationary phase. Interactions of ACN with residual silanols are rather weak, and therefore, silanols are uncovered, and water molecules can adsorb on them. The water adsorption on the residual silanols increases with the decreasing amount of water molecules in the mobile phase [10]. The negative part of the excess isotherm is smaller for end-capped stationary phase than for the stationary phase with the same coverage density without end-capping [11,18,63]. An example is provided in Figure 3.9. In this case, the shape of the excess isotherm informs about the presence of accessible residual silanols, which can have a negative influence on the separation process [5,11,103]. It has to be emphasized that negative part of the isotherm informs about adsorption of water, but it has only qualitative meaning. To calculate the amount of adsorbed water, it is necessary to measure excess isotherm of water; however, on RP stationary phases, in most cases, it is impossible because water is a weaker solvent in mixtures with MeOH or ACN in RP conditions.

The case of methanol-water system is more complicated than of ACN-water system because of the possibility of MeOH for interaction by forming hydrogen bonds. The adsorption of methanol on RP stationary phases is governed by both mechanisms—hydrophobic effect and polar interaction with residual silanols [18]. Therefore, the low-coverage-density phase is better solvated by methanol molecules than by ACN, which can solvate only organic ligands and cannot interact with residual silanols.

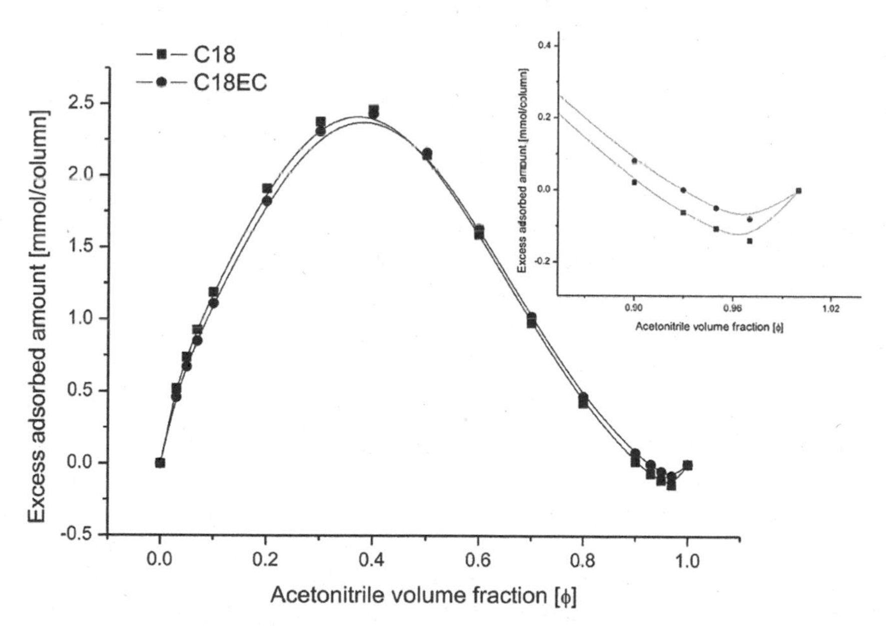


**FIGURE 3.9** Excess isotherms of ACN on end-capped (EC) and non-end-capped C18 stationary phase, together with enlarged negative part of the isotherms.

In ACN–water mobile phase, the residual silanols are covered by water molecules, whereas in methanol–water system, the competition occurs between methanol and water molecules. It indicates that if the coverage density is low enough for methanol molecules to penetrate between bonded ligands to the silica surface, their interaction with silanols is stronger than the interaction between silanols and water molecules. As a result of such interactions, the excess isotherms of MeOH usually do not behave the negative part [18,63].

### 3.5.4 Physical Parameters

Physical parameters, the temperature and the pressure, have considerable impact on chromatographic separations [107]. These parameters influence the retention of the solutes and solvation processes of the stationary phase by solvent molecules from the mobile phase [108]. It is expected that the temperature and pressure have an impact on the distribution of the solvents from the mobile phase to the stationary bonded phases [109,110].

#### 3.5.4.1 Temperature

The influence of temperature on solvent adsorption was measured for two solvents mixtures. For ACN-water conditions, it was investigated by Kazakevich and McNair [5] and, for methanol-water mobile phase, by Poplewska et al. [96]. As expected, the decrease of solvent adsorption was observed with the increase of temperature. These changes are better visible for stronger solvent—ACN, which has relatively higher adsorption [95,111]. Similar results were also obtained for Ultra-High Performance Liquid Chromatography (UHPLC) system using core-shell columns [112]. Exemplary results are presented in Figure 3.10.

The impact of temperature on solvent adsorption in UHPLC is limited. In HPLC system, a temperature increase from 293 to 323 K causes the decrease of ACN adsorption of about 40% of the initial value. In UHPLC system, this decrease is much lower, less than 25% [112]. It has to be emphasized that the temperature range in UHPLC measurement has to be understood as a temperature inside the column thermostat. Temperature inside UHPLC column may vary from set temperature, as was proven by D. Åsberg et al. [113].

It may be concluded that considerable influence of temperature on solvent adsorption is observed when the modifier strongly solvates the stationary phase. The decrease of solvent adsorption will result in the decreasing of the elution strength.

#### 3.5.4.2 Pressure

The influence of pressure on solvation processes may be illustrated in comparison between the traditional HPLC system and UHPLC. In HPLC, surface chemistry has a significant influence on the preferential adsorption of solvents. One can observe a significant difference between C18 and phenyl-bonded or other stationary phases (see part 3.5.2). Preferential adsorption of ACN on phenyl columns is almost two times higher than that on alkyl C18 stationary phase (Figure 3.8). It confirms that

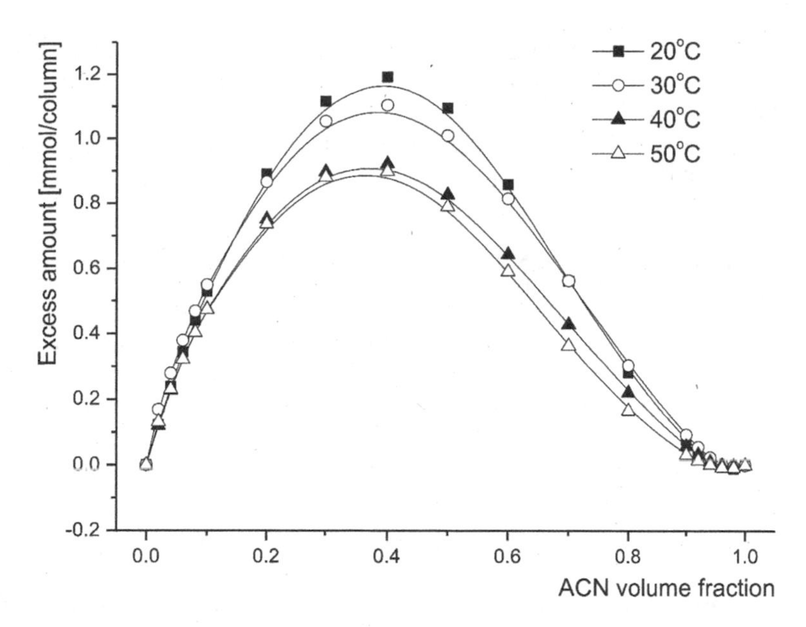


**FIGURE 3.10** Temperature influence on acetonitrile adsorption. (Based on Bocian, S. et al., *J. AOAC. Int.*, 100, 1647–1651, 2017.)

in HPLC, functional groups present on the silica gel surface (or on other support) decide about the preferential adsorption of solvents.

In the case of UHPLC, all registered excess adsorption isotherms exhibit similar shape and similar values. There are only small differences between the maximum of excess isotherms. In the results presented by Bocian et al. [112], maximum of excess isotherms varies from around 0.9 mmol/column to 1.2 mmol/column, which suggests no preferential adsorption on phenyl-hexyl and biphenyl stationary phases in comparison with C18 stationary phase.

Comparing HPLC with UHPLC system during excess isotherm measurement, the backpressure was around 8 times lower in HPLC system. In addition, during excess isotherm determination, the mobile phase composition is changed from pure organic modifier to pure water. As a result, the pressure at each point of the isotherms is different, with the maximum around 40–60% of organic modifier in water. In this case, direct discussion about the pressure in such experiment is impossible. However, it was approximated that, in HPLC using home-made stationary phases, the average pressure was around 60 bars and, in UHPLC system with Kinetex columns, it was in the range 300–400 bar [112]. It means that the pressure increase is significant and that its influence on solvent adsorption is meaningful.

The only possible reason of the observed phenomena is that higher pressure in UHPLC presses the molecule from the mobile phase to the surface of the stationary phase. Pressure impact is more favorable than preferential adsorption of solvent owing to specific and nonspecific interactions. In UHPLC, system solvation process is somehow unified by the pressure, and solvation processes are almost independent of the specific surface properties, or at least the role of intermolecular interactions is significantly lower. It is in opposition to HPLC conditions where specific interactions differentiate solvation processes [112].

## 3.6 THERMODYNAMICS OF SOLVATION PROCESSES

In liquid chromatography, the retention of solute is governed by partitioning between the layers of bonded nonpolar groups in the stationary phase or by adsorption or their combination [4]. The partition coefficients can be transformed to free energy ($\Delta G$) of transfer involving enthalpic ($\Delta H$) and entropic ($\Delta S$) components:

$$\Delta G = \Delta H - T\Delta S$$

These components may be experimentally determined by calorimetric measurement, which has been reported frequently in the literature [114]. The relationship between the free energy, retention factor, $k$, and thermodynamic temperature, $T$ (in Kelvin), is described according to the following expression (van't Hoff equation) [43,115]:

$$\ln k_i = \ln K + \ln\frac{V_S}{V_M} = \frac{-\Delta G^0}{RT} + \ln\frac{V_S}{V_M} = \frac{\Delta S^0}{R} + \ln\frac{V_S}{V_M} - \frac{\Delta H^0}{RT} = A_i + \left(\frac{B_i}{T}\right) \quad (3.9)$$

The dependence of ln $k$ versus $T^{-1}$ plots should embody linearity [116–118]. The parameter $B_i$ involves the standard partial molar enthalpy of the solute $i$ transfer from the mobile phase to the stationary phase, $-\Delta H^0$. The parameter $A_i$ is proportional to the standard partial molar entropy of the solute transfer from the mobile phase to the stationary phase, $\Delta S_0$, and includes the phase ratio (the ratio of the volumes of the stationary, $V_S$, and of the mobile, $V_M$) in the chromatographic system [119]. The $V_M$ is the column hold-up volume essential for the determination of the retention factor, $k$. $R$ is the gas constant, and $T$ is the thermodynamic temperature in Kelvin [120–122]. By plotting ln $k$ versus $T^{-1}$ over a sufficiently broad temperature range, the enthalpic and the entropic contributions to retention, $-\Delta H^0$ from the slope, and $\Delta S^0$ from the intercept of the plot may be calculated. The same procedure may be used for the measurement of solvent molecules interactions with the stationary phase [111].

Van't Hoff plots can provide the information on whether or not the retention mechanisms change over the studied temperature range [123,124]. It has been acknowledged that phase transition phenomenon may cause some small deviations from linearity [125,126]. That is the consequence of a change in the molecular structure of the stationary phase, which appears in the 20°C–50°C range for C18 silica-based stationary phases.

In the case of the solute, the standard enthalpy of the retention in RPLC (i.e., a solute transfer from the mobile phase to the stationary phase) is favorable, owing to the large and favorable stationary-phase contribution, which actually overcomes an unfavorable mobile-phase contribution to the enthalpy of retention [127,128]. The net free energy of retention is favorable, owing to the favorable enthalpic contribution to retention, which arises from the net interactions in the stationary phase. Entropic contributions to retention are not controlling. In RPLC, retention is because of the enthalpically dominated lipophilic interaction of nonpolar solutes with the stationary phase and not from solvophobic processes in the mobile phase [127]. These results support a "partition-like" mechanism of retention rather than an "adsorption-like" mechanism. It confirms that the stationary phase plays a significant role in the overall retention process.

In the case of solvation processes on the stationary phases, it is commonly known that solvents exhibit weak adsorption on the stationary phase in comparison with most of the organic compounds. Solvents are usually eluted in the void volume of the chromatographic column. However, in RPLC system at high water concentration in the binary mobile phase, a significant retention of organic solvent may be found.

Adsorption of methanol from water solution exhibits higher enthalpy than in the case of ACN, which may be attributed to the possibility of hydrogen bond creation. It was confirmed by chromatographic measurement [111] and microcalorimetry [83]. From among the four solvents (MeOH, EtOH, 2-PrOH, and ACN), the highest values of enthalpy ($-\Delta H$) was observed for methanol—the smallest and the most polar molecule. The lowest enthalpy was measured for propan-2-ol, which is the biggest molecule and the most hydrophobic from the tested group. It suggests that the solvent adsorption enthalpy depends on the polar interaction with residual silanols and on the possibility of penetration between the bonded ligands of the stationary phase, which depends on the size of the molecule [111].

The solvent adsorption process is exothermal. The enthalpy increases (gives more negative values) with the increase of the organic solvent concentration in the mobile phase. It is a result of competitive adsorption of water and organic modifiers on the polar residual silanols near the support surface [111]. When water molecules block most of the polar adsorption sites, the enthalpy of solvent adsorption decreases because organic solvent molecules adsorbs mostly on bonded ligands owing to weak dispersive forces. It was also confirmed by microcalorimetric measurements published by Buszewski et al. [76], where water-methanol mixtures provide lower thermal effect (enthalpy) of stationary phase immersion than pure methanol.

Enthalpy of solvation changes with the coverage density of hydrophobic stationary phase in parabolic manner. It may be a result of two different phenomena. First, there are the polar interactions with residual silanols, when the coverage of the bonded ligands is low. If the concentration of bonded ligands increases, the number of residual silanols decreases. As a result, the enthalpy decreases as well. However, further increase of the surface coverage density causes the formation of dense bonded layer. In that case, the most dominant process taking place is the solvation of huge number of octadecyl ligands via dispersive interactions. As a result, the enthalpy increases with the increased number of bonded ligands on the stationary-phase surface. Similar parabolic trend is observed for entropy [111].

The entropy of adsorption for MeOH and ACN reaches mostly negative values, whereas for EtOH and 2-PrOH, the positive values are observed. In the case of the enthalpy changes, the positive values were observed only for propan-2-ol. A positive value of entropy changes indicates a decrease in order of the chromatographic system, as the solvent molecule is transferred from the mobile phase to the stationary phase, which is the evidence of the hydrophobic effect [129].

The calculation of Gibbs free energy ($\Delta G$) provides the negative values for ACN, whereas for MeOH, it provides positive values. The negative values of ACN $\Delta G$ confirm the adsorption process on the stationary phase. The positive values of $\Delta G$ in the case of MeOH are against the adsorption process. However, it has to be remembered that chromatographic experiments with thermodynamic parameters' determination for solvent adsorption are carried out at high water content in the mobile phase. As it was proven by Bocian et al. [18], adsorption of MeOH and water is a competitive phenomenon. Adsorption of MeOH is also much weaker than that of ACN. For the extremely low covered stationary phase, the $\Delta G$ of ACN is also positive, which may confirm the theory of significant adsorption of water on the residual silanols and the displacement effect of organic modifier by the adsorption of water.

## 3.7 SOLVATION ON POLAR-EMBEDDED STATIONARY PHASES

The presence of polar functional groups changes the nature of the surface. Polar functional groups create adsorption center for interaction with polar molecules. In such case, polar groups are better solvated by water or methanol than by ACN. Usually, it is a result of hydrogen bond formation between water or MeOH and functional groups that contain nitrogen or oxygen atoms. However, dipole-type interactions of polar groups with ACN are also possible.

In the previous part, it was proven that phenyl-bonded phases preferentially adsorb ACN in comparison with MeOH as a result of $\pi$-electrons interaction [12]. Interesting observation was made on phenyl-bonded phases with incorporated polar groups. In the case of phenyl polar embedded stationary phases, all of them possess a phenyl ring that preferentially adsorbs ACN molecules. However, the presence of polar groups increases the water adsorption and reduces the adsorption of ACN. It is illustrated in Figure 3.11.

In addition, the differences are observed in relative values of ACN and MeOH adsorption. Preferential adsorption of MeOH is observed in the case of material with amine group via hydrogen bonding and various polar interactions [130]. Phenyl-bonded phase with amine group adsorbs more polar solvent, MeOH, or water than typical phenyl-bonded phase. Region of polar interactions is significantly greater than in the case of typical RP-bonded phases. It is illustrated in Figure 3.12. It also has effect on the relative elution strength of the solvent and on the retention properties.

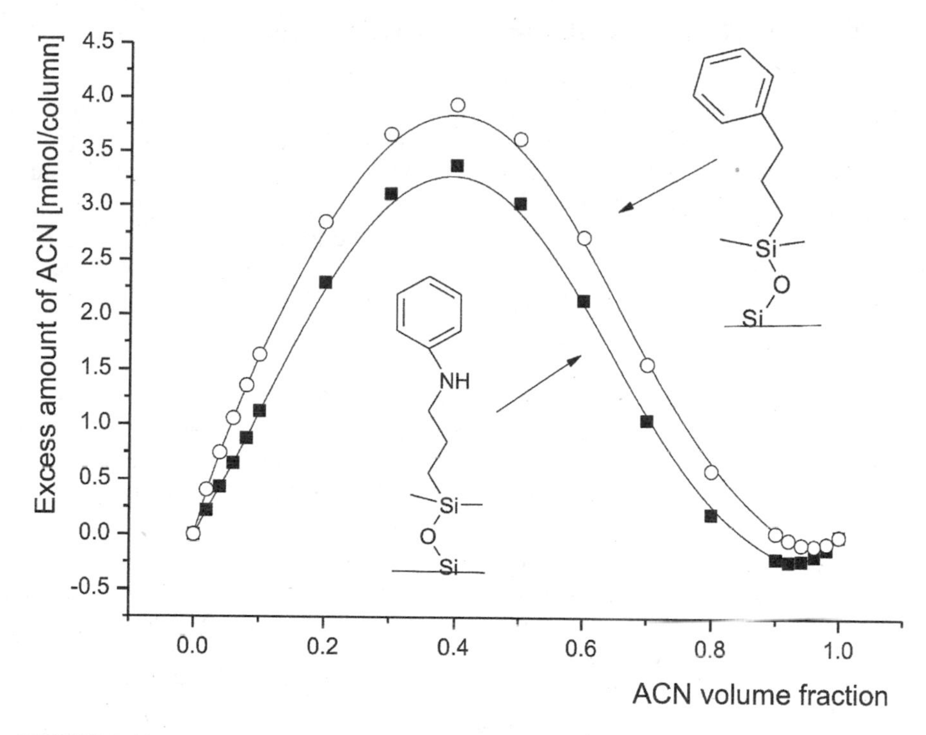


FIGURE 3.11 Influence of polar functional groups on ACN adsorption.

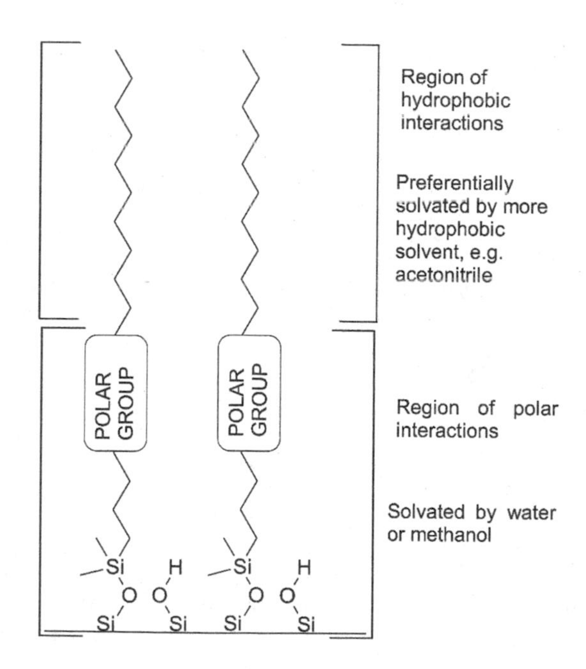


FIGURE 3.12 Solvation region of polar embedded RP stationary phase.

## 3.8 COMPETITIVENESS OF SOLVENT ADSORPTION

Surface of the stationary phase is heterogeneous [48]. If the coverage of bonded ligands on the stationary-phase surface is high, organic modifier molecules cannot penetrate the bonded ligands. They adsorb only on the top of the ligand. However, smaller molecules of water can penetrate the ligands to the silica gel surface and interact with residual silanols. Consequently, the preferential adsorption of water may be observed as a negative part of the organic solvent excess isotherm [17]. For the lowest-coverage organic solvent (e.g., methanol), molecules can penetrate between ligands to the silica surface and interact with the residual accessible silanols. The competitive adsorption of both solvents (methanol and water) on the residual silanols is observed.

Competiveness of adsorption is better visible in HILIC. Each time, negative part of excess isotherm of one solvent corresponds to positive part of isotherm of the second solvent. When ACN is preferentially adsorbed, the water molecules are displaced, and opposite, preferential adsorption of water displaces ACN molecules from the stationary-phase surface. What is interesting is that, in the constant pore volume of the packing material, the volume of excessively adsorbed water equals the volume of displaced ACN at ACN-rich mobile phase and the volume of adsorbed ACN equals the volume of displaced water in water-rich mobile phase. Measurements of water and ACN excess isotherms also give the answer about the potential application of given adsorbent in RP and HILIC conditions. Significant adsorption of ACN confirms the application of a given material in RP condition, whereas preferential adsorption of water classifies adsorbent to HILIC [15].

It is commonly known that the adsorption of solvent in chromatographic systems is a competitive phenomenon. Depending on the surface properties, both solvents, organic and water, may adsorb on the surface. In RPLC system, where stationary phases are hydrophobic, organic solvents adsorb preferentially, owing to stronger affinity to the hydrophobic surface, and a more hydrophobic organic solvent adsorbs more strongly than a more polar organic solvent. In contrast, polar molecules of water preferentially solvate polar functional groups, such as residual silanols in RPLC, as well as various hydroxyl groups, amine groups, etc., in HILIC.

The presence of an amine group in the structure of a phenyl-bonded ligand significantly influences the polarity of the stationary phase and changes the solvation processes in binary hydro-organic mobile phase. The requirement for excess adsorption determination is that the isotherm of the stronger adsorbed solvent may be determined from the solution of weakly adsorbed solvent. In the case of stationary phases with hydrophobic and polar functional groups, the measurements of both solvents from binary solution are possible if both solvents adsorb significantly on the stationary-phase surface [101].

In a series of various phenyk-bonded phases, the excess adsorption of water was determined on the phenyl-amine stationary phase. On the other phenyl-bonded stationary phases, the measurement of water adsorption was impossible, owing to the weak adsorption of water [101]. In Figure 3.13, excess isotherms of ACN from water and of water from ACN are plotted. Both isotherms are plotted as a function of ACN volume fraction, so that the water excess isotherm is presented in the reversed direction.

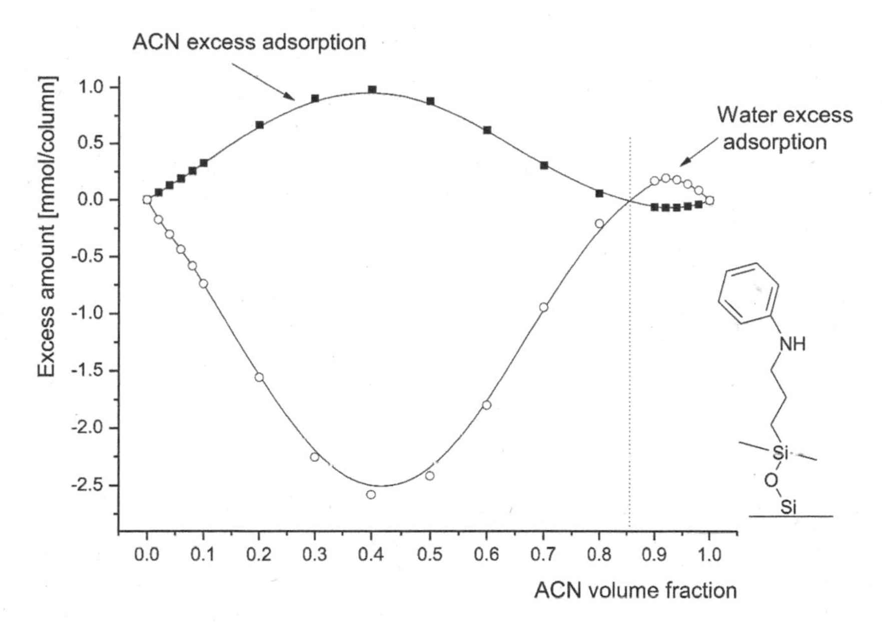


**FIGURE 3.13** Excess adsorption isotherms of water and ACN on phenyl-amine stationary phase.

In Figure 3.13, the excess adsorption of ACN corresponds to a negative part of the water adsorption isotherm, confirming the competitiveness of ACN and water adsorption. It is an ideal confirmation of the displacement effect that occurs in a liquid chromatographic system. Depending on the concentration of a given solvent in the binary mobile phase, the adsorption strength varies. At low concentration of ACN in the mobile phase, the preferential adsorption of this solvent is observed. As a result, the concentration of water at the mobile-phase/stationary-phase interface is lower than in the bulk mobile phase. When the concentration of ACN increases (higher than 85% v/v), the excess adsorption of water may be observed owing to the adsorption of water molecules on the polar functionalities [58,11]. The adsorption of water is observed as a negative part of the excess isotherm plotted for a given organic solvent. In the case of water excess isotherm, when the concentration of water at the interface is higher than in the bulk mobile phase, it is visible as a positive part of the water excess isotherm at the high concentration of ACN in the mobile phase. In addition, the mobile phase composition may be found, at which the compositions of the adsorbed layer of solvents and the mobile phase are equal. In the case of the phenyl-amine stationary phase in an ACN-water mobile phase, such a composition is equal to 85% ACN.

## 3.9 INFLUENCE OF SOLVATION ON CHROMATOGRAPHIC ANALYSES

The solvation of a given stationary phase is different in the MeOH-water and ACN-water environments. These differences depend on the type of the stationary phase. Depending on the functional group present on the stationary-phase surface, some stationary phases preferentially adsorb methanol and other phases preferentially adsorb ACN as a result of intermolecular interaction between particular functional groups in the stationary phase and functional groups of the solvent molecule. If organic solvent adsorbs on the stationary phase more strongly, its elution strength increases. Thus, it can be concluded that the elution solvent strength is a function of solvent adsorption. Stronger adsorption of a given solvent causes higher elution strength [9,33,38,101].

When stationary phases preferentially adsorb MeOH in comparison with ACN, the elution strength of MeOH should be relatively stronger. This should result in lower retention of analyzed compounds. On the other hand, when ACN is preferentially adsorbed, its elution strength is relatively higher than that of MeOH. This effect may be proved in the case of theoretically iso-eluotropic mobile phases.

In Figure 3.14, the results of the retention measurements of benzene, naphthalene, and phenanthrene are presented. Measurements were carried out in iso-eluotropic mobile phases MeOH/water 60/40$_{v/v}$ and ACN/water 50/50$_{v/v}$. Those conditions were chosen based on the nomogram (iso-eluogram). In such conditions, on C18 stationary phases, retention times of solute tested should be the same. However, different situation is observed in the case of phenyl-bonded materials. On the phenyl-amine

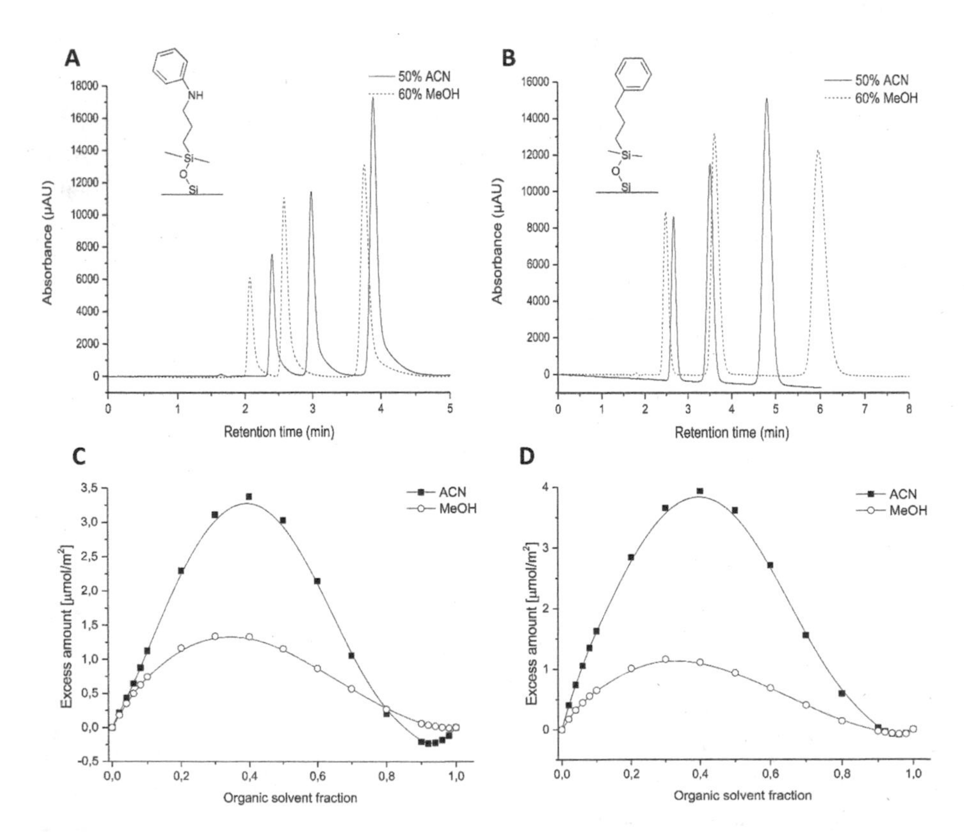


**FIGURE 3.14** Separation of benzene, naphthalene, and phenanthrene and excess isotherms of MeOH and ACN on phenyl-amine (A and C) and phenyl-propyl (B and D) stationary phase.

stationary phase, the retention of all solutes in MeOH-water mixture is lower than in the ACN-water mobile phase. It confirms that on this stationary phase, adsorption of MeOH is relatively stronger, and thus, the elution strength of MeOH is higher. The opposite situation is observed on phenyl-propyl stationary phase. Despite isoeluotropic mobile phases, retention in the MeOH-water mixture is much higher in comparison with the mobile phase composed of ACN and water. It is a result of preferential adsorption of ACN molecules when phenyl-bonded stationary phases do not contain any polar groups to enable polar interactions with MeOH.

## 3.10 INFLUENCE OF SOLVATION ON COLUMN VOID VOLUME

Colum void volume is a critical parameter in liquid chromatography. Most theoretical relationships in chromatographic theory deal with the retention factor, and the accuracy of the void volume determination plays an important role in all calculations [53,57]. In the basic assumptions, void volume of the column is considered constant in any eluent type and composition of the eluent. However, it is observed that elution volume of "unretained marker" depends on the qualitative and quantitative mobile-phase composition. This indicates the exclusion of the "unretained marker" molecules from the adsorbent surface and preferential adsorption of the organic eluent components. Preferential adsorption of organic solvents is illustrated in Figure 3.15, where elution volumes of various organic solvents are plotted over their concentration in water mixtures.

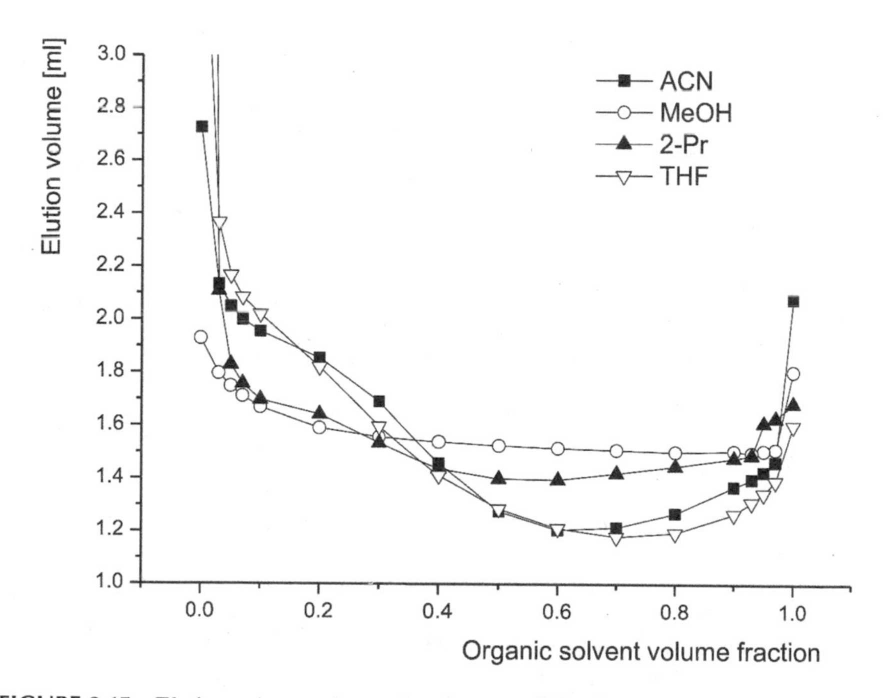


**FIGURE 3.15** Elution volumes of organic solvent on C18 column.

It has to be emphasized that the highest differences are observed for high water content in the mobile phase; however, thermodynamic void volumes calculated according to Equation 3 are consistent [53]. Nevertheless, the structure and the conformation of the stationary phase depend on the solvation processes. The structure of chemically bonded ligands changes with the qualitative and quantitative composition of the mobile phase, which results in the changes of the interface area between the bonded phase and bulk mobile phase.

## 3.11 CONCLUSIONS

Solvation process is an important part of chromatographic elution. As a result of solvent molecules attraction by the surface, the stationary phase is solvated by the molecules from the mobile phase. Solvent molecules preferentially solvate active centers of the stationary phase, depending on its character. Polar adsorption sites are preferentially solvated by water or protic solvents such as alcohols, and hydrophobic centers are solvated by more hydrophobic molecules. As a result, in RP system, hydrophobic stationary phase preferentially adsorbs organic solvent, whereas polar stationary phase in HILIC mode preferentially adsorbs water molecules. Adsorption of water on residual silanols may create a hydrophilic pillow near the support surface. Thus, solvation processes that take place in chromatographic column depend on many parameters of the stationary phase.

Measurement of solvation process may be used for the characterization of hydrophobic and hydrophilic properties of the stationary phase. On the other hand, the knowledge about the solvation processes may be utilized in the development of new chromatographic packings. The knowledge of solvation process is crucial for understanding the retention mechanism in various modes of liquid chromatography as well as for the optimization of chromatographic separation in routine analyses. The control of the retention and the selectivity of the separation that is performed by changing the composition of the mobile phase or changing the type of organic modifier are a play with solvation processes that take place in chromatographic column.

## REFERENCES

1. Sandi, A., and L. Szepesy. 1999. Evaluation and modulation of selectivity in reversed-phase high-performance liquid chromatography. *J. Chromatogr. A* 845:113–131.
2. Gritti, F., and G. Guiochon. 2005. Influence of the pressure on the properties of chromatographic columns III. Retention volume of thiourea, hold-up volume, and compressibility of the C18-bonded layer. *J. Chromatogr. A* 1075:117–126.
3. Riedo, F., and E. Kovats. 1982. Adsorption from liquid mixtures and liquid chromatography. *J. Chromatogr.* 239:1–28.
4. Gritti, F., Y. V. Kazakevich, and G. Guiochon. 2007. Effect of the surface coverage of endcapped C18-silica on the excess adsorption isotherms of commonly used organic solvents from water in reversed phase liquid chromatography. *J. Chromatogr. A* 1169:111–124.
5. Kazakevich, Y. V., and H. M. McNair. 1995. Study of the excess adsorption of the eluent components on different reversed-phase adsorbents. *J. Chromatogr. Sci.* 33:321–327.

6. Eletkov, A., and V. Kazakievich. 1987. Comparison of various chromatographic methods for the determination of adsorption isotherms in solutions. *J. Chromatogr.* 395:473–480.
7. Jaroniec, M. 1993. Partition and displacement models in reversed-phase liquid chromatography with mixed eluents. *J. Chromatogr. A* 656:37–50.
8. Kazakevich, Y. V. 2006. High-performance liquid chromatography retention mechanisms and their mathematical descriptions. *J. Chromatogr. A* 1126:232–243.
9. McCormick, R. M., and B. L. Karger. 1980. Distribution phenomena of mobile-phase components and determination of dead volume in reversed-phase liquid chromatography. *Anal. Chem.* 52:2249–2257.
10. Bocian, S., A. Felinger, and B. Buszewski. 2008. Comparison of solvent adsorption on chemically bonded stationary phases in RP-LC. *Chromatographia* 68:S19–S26.
11. Buszewski, B., S. Z. Bocian, and A. Felinger. 2008. Excess isotherms as a new way for characterization of the columns for reversed-phase liquid chromatography. *J. Chromatogr. A* 1191:72–77.
12. Bocian, S., P. Vajda, A. Felinger, and B. Buszewski. 2008. Solvent excess adsorption on the stationary phases for reversed-phase liquid chromatography with polar functional groups. *J. Chromatogr. A* 1204:35–41.
13. Buszewski, B., J. Schmid, K. Albert, and E. Bayer. 1991. Chemically bonded phases for the reversed-phase high-performance liquid chromatographic separation of basic substances. *J. Chromatogr.* 552:415–427.
14. Gadzała-Kopciuch, R., and B. Buszewski. 2003. A comparative study of hydrophobicity of octadecyl and alkylamide bonded phases based on methylene selectivity. *J. Sep. Sci.* 26:1273–1283.
15. Noga, S., S. Bocian, and B. Buszewski. 2013. Hydrophilic interaction liquid chromatography columns classification by effect of solvation and chemometric methods. *J. Chromatogr. A* 1278:89–97.
16. Buszewski, B., and S. Noga. 2012. Hydrophilic interaction liquid chromatography (HILIC)—A powerful separation technique. *Anal. Bioanal. Chem.* 402:231–247.
17. Bocian, S. Z., P. Vajda, A. Felinger, and B. Buszewski. 2009. Excess adsorption of commonly used organic solvents from water on nonend-capped C18-bonded phases in reversed-phase liquid chromatography. *Anal. Chem.* 81:6334–6346.
18. Bocian, S. Z., P. Vajda, A. Felinger, and B. Buszewski. 2010. Effect of end-capping and surface coverage on the mechanism of solvent adsorption. *Chromatographia* 71:S5–S11.
19. Glenne, E., K. Öhlén, H. Leek, M. Klarqvist, J. Samuelsson, and T. Fornstedt. 2016. A closer study of methanol adsorption and its impact on solute retentions in supercritical fluid chromatography. *J. Chromatogr. A* 1442:129–139.
20. Vajda, P., and G. Guiochon. 2013. Surface excess isotherms of organic solvent mixtures in a system made of liquid carbon dioxide and a silicagel surface. *J. Chromatogr. A* 1308:139–143.
21. Vajda, P., and G. Guiochon. 2013. Modifier adsorption in supercritical fluid chromatography onto silica surface. *J. Chromatogr. A* 1305:293–299.
22. Enmark, M., P. Forssén, J. Samuelsson, and T. Fornstedt. 2013. Determination of adsorption isotherms in supercritical fluid chromatography. *J. Chromatogr. A* 1312:124–133.
23. Vajda, P., A. Felinger, and A. Cavazzini. 2010. Adsorption equilibria of proline in hydrophilic interaction chromatography. *J. Chromatog. A* 1217:5965–5970.
24. Alpert, A. J. 1990. Hydrophilic-interaction chromatography for the separation of peptides, nucleic acids, and other polar compounds. *J. Chromatogr.,* 449:177–196.
25. Gritti, F., A. dos Santos Pereira, P. Sandra, and G. Guiochon. 2009. Comparison of the adsorption mechanisms of pyridine in hydrophilic interaction chromatography and in reversed-phase aqueous liquid chromatography. *J. Chromatog. A* 1216:8496–8504.

26. Melnikov, S. M., A. Holtzel, A. Seidel-Morgenstern, and U. Tallarek. 2012. A Molecular dynamics study on the partitioning mechanism in hydrophilic interaction chromatography. *Angew. Chem. Int. Ed.* 51:6251–6254.
27. Jandera, P. 2008. Stationary phases for hydrophilic interaction chromatography, their characterization and implementation into multidimensional chromatography concepts. *J. Sep. Sci.* 31:1421–1437.
28. Bocian, S., A. Nowaczyk, and B. Buszewski. 2012. New alkyl-phosphate bonded stationary phases for liquid chromatographic separation of biologically active compounds. *Anal. Bioanal. Chem.* 404:731–740.
29. Bocian, S. Z., and B. Buszewski. 2015. Synthesis and characterization of phosphodiester stationary bonded phases for liquid chromatography. *Talanta* 143:35–41.
30. Cavazzini, A., G. Nadalini, V. Malanchin, V. Costa, F. Dondi, and F. Gasparrini. 2007. Adsorption mechanisms in normal-phase chromatography. Mobile-Phase modifier adsorption from dilute solutions and displacement effect. *Anal. Chem.* 79:3802–3809.
31. Jandera, P., and J. Churacek. 1985. *Gradient Elution in Column Liquid Chromatography: Theory and Practice*. Amsterdam, the Netherlands: Elsevier.
32. Glajch, J. L., and L. R. Snyder. 1990. *Computer-assisted method development for high-performance liquid chromatography*. Amsterdam, the Netherlands: Elsevier Science & Technology.
33. McCormick, R. M., and B. L. Karger. 1980. Role of organic modifier sorption on retention phenomena in reversed-phase liquid chromatography. *J. Chromatogr.* 199:259–273.
34. Findenegg, G. H., and F. Koster. 1986. A new equation for the retention of solutes in liquid–solid adsorption chromatography with mixed mobile phases. *J. Chem. Soc., Faraday Trans.* 82:2691–2705.
35. Katz, E. D., K. Ogan, and R. P. Scott. 1986. Distribution of a solute between two phases: The basic theory and its application to the prediction of chromatographic retention. *J. Chromatogr.* 352:67–90.
36. Hafkenscheid, T. L., and E. Tomlinson. 1986. Estimation of physicochemical properties of organic solutes using HPLC retention parameters. *Adv. Chromatogr.* 25:1–62.
37. Everett, D. H. 1964. Thermodynamics of adsorption from solutions I. *J. Chem. Soc., Faraday Trans.* 60:1803–1813.
38. Ha, N. L., J. Ungvarai, and E. Kovats. 1982. Adsorption isotherm at the liquid-solid interface and the interpretation of chromatographic data. *Anal. Chem.* 54:2410.
39. Knox, J. H., R. Kaliszan, and G. Kennedy. 1980. Enthalpic exclusion chromatography. *Faraday Symp. Chem. Soc.* 15:113.
40. Wang, M., J. Mallette, and J. F. Parcher. 2008. Strategies for the determination of the volume and composition of the stationary phase in reversed-phase liquid chromatography. *J. Chromatogr. A* 1190:1–7.
41. Guiochon, G., A. Felinger, D. G. Shirazi, and A. Katti. 2006. *Fundamentals of Preparative and Nonlinear Chromatography*. Amsterdam, the Netherland: Elsevier Academic Press.
42. Eltekov, Y. A., and Y. V. Kazakevich. 1976. Investigation of adsorption equilibrium in chromatographic columns by the frontal method. *J. Chromatogr.* 365:213–219.
43. Melander, W. R., and C. S. Horvath. 1980. In *High-Performance Liquid Chromatography, Advances and Perspectives Vol. 2*, Edited by C. Horvath. New York: Academic Press.
44. Vajda, P., A. Felinger, and G. Guiochon. 2013. Evaluation of surface excess isotherms in liquid chromatography. *J. Chromatogr. A* 1291:41–47.
45. Köster, F., and G. H. Findenegg. 1982. Adsorption from binary solvent mixtures onto silica gel by HPLC frontal analysis. *Chromatographia* 15:743–747.

46. Soukup, J., and P. Jandera. 2014. Adsorption of water from aqueous acetonitrile on silica-based stationary phases in aqueous normal-phase liquid chromatography. *J. Chromatogr. A* 1374:102–111.
47. Soukup, J., P. Janas, and P. Jandera. 2013. Gradient elution in aqueous normal-phase liquid chromatography on hydrosilated silica-based stationary phases. *J. Chromatogr. A* 1268:111–118.
48. Vajda, P., S. Bocian, B. Buszewski, and A. Felinger. 2010. Examination of the surface heterogeneity of reversed-phase packing materials with solvent adsorption. *J. Sep. Sci.* 33:3644–3654.
49. Ohashi, J., M. Harada, and T. Okada. 2017. The application of the minor disturbance method is limited to the determination of excess adsorption isotherms of electrolyte-free binary mixtures. *J. Sep. Sci.* 40:842–848.
50. Foti, G., C. De Reyff, and E. S. Kovats. 1990. Method of chromatographic determination of excess adsorption from binary liquid mixtures. *Langmuir* 6:759–766.
51. Wang, H. L., J. L. Duda, and C. J. Radke. 1978. Solution adsorption from liquid chromatography. *J. Colloid Interface Sci.* 66:153–165.
52. Kazakevich, Y. V., and H. M. McNair. 1993. Thermodynamic definition of HPLC dead volume. *J. Chromatogr. Sci.* 31:317–322.
53. Kazakevich, Y. V., and R. LoBrutto. 2007. *HPLC for Pharmaceutical Scientists*. Hoboken, NJ: John Wiley & Sons.
54. Gritti, F., and G. Guiochon. 2005. Adsorption mechanism in RPLC. Effect of the nature of the organic modifier. *Anal. Chem.* 77:4257–4272.
55. Wang, M., J. Mallette, and J. F. Parcher. 2009. Sorption isotherms of ternary eluents in reversed-phase liquid chromatography. *Anal. Chem.* 81:984–990.
56. Buszewski, B., S. Bocian, and A. Felinger. 2012. Artifacts in liquid-phase separations–system, solvent, and impurity peaks. *Chem. Rev.* 112:2629–2641.
57. Knox, J. H., and R. Kaliszan. 1985. Theory of solvent disturbance peaks and experimentaldetermination of thermodynamic dead-volume in column liquid chromatography. *J. Chromatogr.* 349:211–234.
58. Kazakevich, Y. V., R. LoBrutto, F. Chan, and T. Patel. 2001. Interpretation of the excess adsorption isotherms of organic eluent components on the surface of reversed-phase adsorbents Effect on the analyte retention. *J. Chromatogr. A* 913:75–87.
59. Rustamov, I., T. Farcas, F. Ahmed, F. Chan, R. LoBrutto, H. M. McNair, and Y. V. Kazakevich. 2001. Geometry of chemically modified silica. *J. Chromatogr. A* 913:49–63.
60. Kirkland, J. J., M. A. van Straten, and H. A. Claessens. 1995. High pH mobile phase effects on silica-based reversed-phase high-performance liquid chromatographic columns. *J. Chromatogr. A* 691:3–19.
61. Hayrapetyan, S. S., U. D. Neue, and H. G. Khachatryan. 2006. Changes in the pore character of silicas caused by surface bonding and polymer coating. *J. Sep. Sci.* 29:810–819.
62. Slaats, E. H., W. Markowski, J. Fekete, and H. Pope. 1981. Distribution equilibria of solvent components in reversed-phase liquid chromatographic columns and relationship with the mobile phase volume. *J. Chromatogr.* 207:299–323.
63. Poplewska, I., and D. Antos. 2005. Effect of adsorption of organic solvents on the band profiles in reversed-phase non-linear chromatography. *Chem. Eng. Sci.* 60:1411–1427.
64. Rimmer, C. A., L. C. Sander, S. A. Wise, and J. G. Dorsey. 2003. Synthesis and characterization of C13 to C18 stationary phases by monomeric, solution polymerized, and surface polymerized approaches. *J. Chromatogr. A* 1007:11–20.
65. Giaquinto, A., Z. Liu, A. Bach, and Y. Kazakevich. 2008. Surface area of reversed-phase HPLC columns. *Anal. Chem.* 80:6358–6364.

66. Bass, J. L., B. W. Sands, and P. W. Bratt. 1986. Characterization of bonded silica gel using sorption and thermal analysis. In *Paper Read at Silanes, Surfaces and Interfaces*, p. 267. New York: Gordon and Breach Science Publishers.
67. Kitahara, S., K. Tanaka, T. Sakata, and H. Muraishi. 1981. Porosity change of silica gels by the alkoxylation of their surfaces. *J. Colloid Interf. Sci.* 84:519–525.
68. Bass, J. L., B. W. Sands, and P. W. Bratt. 1985. In D. E. Leyden (Ed.), *Proceedings of the Silanes, Surfaces and Interfaces Symposium*. New York: Gordon and Breach Science Publishers.
69. Sander, L. C., C. J. Glinka, and S.A. Wise. 1985. In D. E. Leyden (Ed.), *Proceedings of the Silanes, Surfaces and Interfaces Symposium*. New York: Gordon and Breach Science Publishers.
70. Bereznitski, Y., M. Jaroniec, and M. E. Gangoda. 1998. Characterization of silica-based octyl phases of different bonding density: Part II. Studies of surface properties and chromatographic selectivity. *J. Chromatogr. A* 828:59–73.
71. Trathnigg, B., M. Veronik, and A. Gorbunov. 2006. Looking inside the pores of a chromatographic column I. Variation of the pore volume with mobile phase composition. *J. Chromatogr. A* 1104:238–244.
72. Gritti, F., and G. Guiochon. 2007. Thermodynamics of adsorption of binary aqueous organic liquid mixtures on a RPLC adsorbent. *J. Chromatogr. A* 1155:85–99.
73. Schay, G. 1969. *In Surface and Colloid Science*: Hoboken, NJ: Wiley-Interscience.
74. Hubbard, A. T., ed. 2002. *Encyclopedia of Surface and Colloid Science*. Volume 1. New York: Marcel Dekker.
75. Buszewski, B., K. Krupczyńska, R. M. Gadzała-Kopciuch, G. Rychlicki, and R. Kaliszan. 2003. Evaluation of HPLC columns: A study on surface homogeneity of chemically bonded stationary phases. *J. Sep. Sci.* 26:313–321.
76. Buszewski, B., K. Krupczyńska, G. Rychlicki, and R. Łobiński. 2006. Effect of coverage density and structure of chemically bonded silica stationary phases on the separation of compounds with various properties. *J. Sep. Sci.* 29:829–836.
77. Staszczuk, P., and B. Buszewski. 1988. Coverage homogenity of chemically bonded C18 phases for HPLC. Investigation of the changes in the free surface energy of the packings. *Chromatographia* 25:881–886.
78. Dubinin, M. M., A. A. Isirikyan, K. M. Nikolaev, N. S. Poiyakov, and L. I. Tatarinova. 1986. Heat of immersion of silica gel in normal alkanes and alcohols. *Russ. Chem. Bull.* 35:1573–9171.
79. Atkins, P. W. 1997. *Physical Chemistry–6th ed.* New York: W. H. Freeman and Company.
80. Buszewski, B., S. Bocian, G. Rychlicki, P. Vajda, and A. Felinger. 2010. Study of solvent adsorption on chemically bonded stationary phases by microcalorimetry and liquid chromatography. *J. Colloid Interf. Sci.* 349:620–625.
81. Bocian, S., G. Rychlicki, M. Matyska, J. Pesek, and B. Buszewski. 2014. Study of hydration process on silica hydride surfaces by microcalorimetry and water adsorption. *J. Colloid Interface Sci.* 416:161–166.
82. Buszewski, B., S. Z. Bocian, and G. Rychlicki. 2011. Investigation of silanol activity on the modified silica surfaces using microcalorimetric measurements. *J. Sep. Sci.* 34:773–779.
83. Buszewski, B., S. Bocian, G. Rychlicki, M. Matyska, and J. Pesek. 2012. Determination of accessible silanols groups on silica gel surfaces using microcalorimetric measurements. *J. Chromatogr. A* 1232:43–46.
84. Samuelsson, J., R. Arnell, J. S. Diesen, J. Tibbelin, A. Paptchikhine, T. Fornstedt, and P. J. R. Sjöberg. 2008. Development of the tracer-pulse method for adsorption studies of analyte mixtures in liquid chromatography utilizing mass spectrometric detection. *Anal. Chem.* 80:2105–2112.

85. Wang, M., B. Avula, Y. H. Wang, J. F Parcher, and I. A. Khan. 2012. Comparison of concentration pulse and tracer pulse chromatography: experimental determination of eluent uptake by bridge-ethylene hybrid ultra-high performance liquid chromatography packings. *J. Chromatogr. A* 1220:75–81.
86. Gritti, F., and G. Guiochon. 2005. Critical contribution of nonlinear chromatography to the understanding of retention mechanism in reversed-phase liquid chromatography. *J. Chromatogr. A* 1099:1–42.
87. Zhang, L., L. Sun, J. I. Siepmann, and M. R. Schure. 2005. Molecular simulation study of the bonded-phase structure in reversed-phase liquid chromatography with neat aqueous solvent. *J. Chromatogr. A* 1079:127–135.
88. Rafferty, J. L., L. Zhang, J. I. Siepmann, and M. R. Schure. 2007. Retention mechanism in reversed-phase liquid chromatography: A molecular perspective. *Anal. Chem.* 79:6551–6558.
89. Rafferty, J. L., J. I. Siepmann, and M. R. Schure. 2008. Influence of bonded-phase coverage in reversed-phase liquid chromatography via molecular simulation I. Effects on chain conformation and interfacial properties. *J. Chromatogr. A* 1204:11–19.
90. Zhang, L., J. L. Rafferty, J. I. Siepmann, B. Chen, and M. R. Schure. 2006. Chain conformation and solvent partitioning in reversed-phase liquid chromatography: Monte Carlo simulations for various water/methanol concentrations. *J. Chromatogr. A* 1126:219–231.
91. Buszewski, B., S. Z. Bocian, and A. Nowaczyk. 2010. Modeling solvation on the chemically modified silica surfaces. *J. Sep. Sci.* 33:2060–2068.
92. Melnikov, S. M., A. Höltzel, A. Seidel-Morgenstern, and U. Tallarek. 2009. Influence of residual silanol groups on solvent and ion distribution at a chemically modified silica surface. *J. Phys. Chem.* 113:9230–9238.
93. Melnikov, S. M., A. Höltzel, A. Seidel-Morgenstern, and U. Tallarek. 2013. Adsorption of water–acetonitrile mixtures to model silica surfaces. *J. Phys. Chem.* 117:6620–6631.
94. Melnikov, S. M., A. Höltzel, A. Seidel-Morgenstern, and U. Tallarek. 2011. Composition, structure, and mobility of water–acetonitrile mixtures in a silica nanopore studied by molecular dynamics simulations. *Anal. Chem.* 83 2569–2575.
95. Buszewski, B., S. Bocian, and R. Zera. 2010. Influence of temperature and pressure on the preferential adsorption of component of hydroorganic mobile phase in liquid chromatography. *Adsorption* 16:437–445.
96. Poplewska, I., W. Piatkowski, and D. Antos. 2006. Effect of temperature on competitive adsorption of the solute and the organic solvent in reversed-phase liquid chromatography. *J. Chromatogr. A* 1103:284–295.
97. Buntz, S., M. Figus, Z. Liu, and Y. V. Kazakevich. 2012. Excess adsorption of binary aqueous organic mixtures on various reversed-phase packing materials. *J. Chromatogr. A* 1240:104–112.
98. Buszewski, B., D. Berek, J. Garaj, I. Novak, and Z. Suprynowicz. 1988. Influence of the porous silica gel structure on the coverage density of a chemically bonded C18 phase for high-performance liquid chromatography. *J. Chromatogr.* 446:191–201.
99. Sands, B. W., Y. S. Kim, and J. L. Bass. 1986. Characterization of bonded-phase silica gels with different pore diameters. *J. Chromatogr.* 360:353–369.
100. Chan, F., L. S. Yeung, R. LoBrutto, and Y. V. Kazakevich. 2005. Characterization of phenyl-type HPLC adsorbents. *J. Chromatogr. A* 1069:217–224.
101. Bocian, S., M. Skoczylas, I. Goryńska, M. Matyska, J. Pesek, and B. Buszewski. 2016. Solvation processes on phenyl-bonded stationary phases: The influence of the polar functional groups. *J. Sep. Sci.* 39:4369–4376.

102. Buszewski, B., M. Jezierska-Świtala, and S. Kowalska. 2003. Stationary phase with specific surface properties for the separation of estradiol diastereoisomers. *J. Chromatogr. B* 792:279–286.
103. McCalley, D. V. 2003. Selection of suitable stationary phases and optimum conditions for their application in the separation of basic compounds by reversed-phase HPLC. *J. Sep. Sci.* 26:187–200.
104. Nawrocki, J. 1997. The silanol group and its role in liquid chromatography. *J. Chromatogr. A* 779:29–71.
105. Buszewski, B., M. Jezierska, M. Wełniak, and D. Berek. 1998. Survey and trends in the preparation of chemically bonded silica phases for liquid chromatographic analysis. *J. High Resolut. Chromatogr.* 21:267–281.
106. Nawrocki, J., and B. Buszewski. 1988. Influence of silica surface chemistry and structure on the properties, structure and coverage of alkyl-bonded phases for high-performance liquid chromatography. *J. Chromatogr.* 449:1–25.
107. Berek, D., and T. Macko. 1989. Pressure effects in high performance liquid chromatography. *Pure Appl. Chem.* 61:2041–2046.
108. Gritti, F., and G. Guiochon. 2006. Adsorption mechanisms and effect of temperature in reversed-phase liquid chromatography. Meaning of the classical Van't Hoff Plot in chromatography. *Anal. Chem.* 78:4642–4653.
109. Berek, D., M. Chalanyova, and T. Macko. 1984. Dependence of preferential solvation of liquid chromatographic sorbents in mixed eluents on pressure *J. Chromatogr.* 286:185–192.
110. Macko, T., M. Chalanyova, and D. Berek. 1986. Effect of Pressure on preferential sorption within column packing: Possible explanation of some unexpected results in GPC with mixed eluents. *J. Liq. Chromatogr.* 9:1123–1140.
111. Bocian, S. Z., J. Soukup, P. Jandera, and B. Buszewski. 2015. Thermodynamics study of solvent adsorption on octadecyl modified silica. *Chromatographia* 78:21–30.
112. Bocian, S., T. Škrinjar, T. Bolanca, and B. Buszewski. 2017. How high pressure unifies solvation processes in liquid chromatography. *J. AOAC. Int.* 100:1647–1651. doi:10.5740/jaoacint.17–0223.
113. Asberg, D., J. Samuelsson, M. Leśko, A. Cavazzini, K. Kaczmarski, and T. Fornstedt. 2015. Method transfer from high-pressure liquid chromatography to ultra-high-pressure liquid chromatography. II. Temperature and pressure effects. *J. Chromatogr. A* 1401:52–59.
114. Ferrannini, E. 1988. The theoretical bases of indirect calorimetry: A review. *Metabolism* 37:287–301.
115. Vanhoenacker, G., and P. Sandra. 2005. High temperature liquid chromatography and liquid chromatography–mass spectroscopy analysis of octylphenol ethoxylates on different stationary phases. *J. Chromatogr. A* 1082:193–202.
116. Li, J., and P. W. Carr. 1997. Effect of temperature on the thermodynamic properties, kinetic performance, and stability of polybutadiene-coated zirconia. *Anal. Chem.* 69:837–843.
117. Takeuchi, T., Y. Watanabe, and D. Ishii. 1981. Role of column temperature in micro high performance liquid chromatography. *J. High Resolut. Chromatogr.* 4:300–302.
118. Greibrokk, T., and T. Andersen. 2003. High-temperature liquid chromatography. *J. Chromatogr. A* 1000:743–755.
119. Škeříková, V., and P. Jandera. 2010. Effects of the operation parameters on hydrophilic interaction liquid chromatography separation of phenolic acids on zwitterionic monolithic capillary columns. *J. Chromatogr. A* 1217:7981–7989.
120. Jandera, P., H. Colin, and G. Guiochon. 1982. Interaction indexes for prediction of retention in reversed-phase liquid chromatography. *Anal. Chem.* 435:435–441.

121. Chester, T. L., and J. W. Coym. 2003. Effect of phase ratio on van't Hoff analysis in reversed-phase liquid chromatography, and phase-ratio-independent estimation of transfer enthalpy. *J. Chromatogr. A* 1003:101–111.
122. Vigh, G., and Z. Varga-Puchony. 1980. Influence of temperature on the retention behaviour of members of homologous series in reversed-phase high-performance liquid chromatography. *J. Chromatogr. A* 196:1–9.
123. Melander, W., D. E. Campbell, and C. S. Horvath. 1978. Enthalpy—entropy compensation in reversed-phase chromatography. *J. Chromatogr.* 158:215–225.
124. Kiridena, W., C. F. Poole, and W. W. Koziol. 2003. Reversed-phase chromatography on a polar endcapped octadecylsiloxane-bonded stationary phase with water as the mobile phase. *Chromatographia* 57:703–707.
125. Nahum, A., and C. Horvath. 1981. Surface silanols in silica-bonded hydrocarbonaceous stationary phases: I. Dual retention mechanism in reversed-phase chromatography. *J. Chromatogr.* 203:53–63.
126. Morel, D., and J. Serpinet. 1981. Influence of the liquid chromatographic mobile phase on the phase transitions of alkyl-bonded silicas studied by gas chromatography. *J. Chromatogr.* 214:202–208.
127. Ranatunga, R. P. J., and P. W. Carr. 2000. A study of the enthalpy and entropy contributions of the stationary phase in reversed-phase liquid chromatography. *Anal. Chem.* 72:5679–5692.
128. Rafferty, J. L., J. I. Siepmann, and M. R. Schure. 2010. Understanding the retention mechanism in reversed-phase liquid chromatography: Insights from molecular simulation. *Adv. Chromatogr.* 48:1–55.
129. Cole, L. A., J. G. Dorsey, and K. A. Dill. 1992. Temperature dependence of retention in reversed-phase liquid chromatography. 2. Mobile-phase considerations. *Anal. Chem.* 64:1324–1327.
130. Bocian, S. Z., and B. Buszewski. 2014. Phenyl-bonded stationary phases-The influence of polar functional groups on retention and selectivity in reversed-phase liquid chromatography. *J. Sep. Sci.* 37:3435–3442.

# 4 Thin-Layer Chromatography in the Determination of Synthetic and Natural Colorants in Foods

*Joseph Sherma*

## CONTENTS

## 4.1 INTRODUCTION

Thin-layer chromatography (TLC) is an important method for the analysis of synthetic and natural colorants in foods, which in this chapter include foodstuffs, beverages, functional foods (nutraceuticals), and dietary supplements. This chapter covers the literature from 2007 to 2018 in order to update three earlier reviews on the analysis of natural colorings in foods by TLC, published in 2007 [1], chromatographic methods for the determination of synthetic food dyes, in which the latest TLC reference cited was published in 2006 [2], and methods for the determination of European Union-permitted natural colors in foods, with one cited TLC reference published in 2009 and all others in 2005 or earlier [3].

TLC is a simple, low-cost, and rapid procedure for the detection of food colorants present in real samples by comparing retardation factor ($R_f$) values and colors of zones in samples with those of standards. It has been proven to be a convenient separation technique that provides accurate and reliable results for purposes such as quality control (QC) of beverages derived from fruits [4] and for screening unknown compounds in complex food matrix [5].

Modern high-performance TLC (HPTLC) has advantages for the identification and quantifications of food colorants and QC of pigment formulations used in printing inks for food packaging, including simplified sample preparation and high robustness with regard to varying matrices owing to single usage of the stationary phase; multidetection methods enabling collection of comprehensive information on a sample; effect-directed analysis (EDA) for profiling of the separated samples by appropriate assays to identify bioactive target and nontarget ingredients; and hyphenation with mass spectrometry (MS), Fourier transform infrared (FTIR) spectrometry, and nuclear magnetic resonance (NMR) spectrometry for the identification and confirmation of sample components [6]. In addition, it is a cost-effective, high-throughput method with low usage of solvents, because multiple samples and standards are chromatographed on a single plate in parallel [7].

Use of chemometrics with TLC is another powerful, recent approach for the analysis of colorants in food products. Colored chromatograms are excellent input data for image and multivariant analysis, for example, for the classification of different foods, wines, and beers [8].

## 4.2 TECHNIQUES OF THIN-LAYER CHROMATOGRAPHY OF FOOD COLORANTS

This section will describe the general techniques that have been applied for food colorant studies by TLC, and the next sections will present examples of selected applications, with specific experimental details provided in each paper.

Colorants are usually extracted from foods prior to TLC by using an appropriate solvent, with traditional techniques such as shake, homogenization, Soxhlet, membrane filtration [9], and ultrasound-assisted extraction (UAE) [10]. Column chromatography may be needed in some cases for removal of interfering coextractives before analysis by TLC, for example, gel permeation [11] and silica gel [12] chromatography. More modern techniques applied for sample preparation prior to TLC analysis include solid-phase extraction (SPE) on octyldecyl-bonded silica gel (C18); polyamide, amino ($NH_2$), and weak-acid polymer-based sorbent (Strata X-AW); and other phases to purify extracts and isolate dye analytes [7]; microwave-assisted extraction [9]; supercritical fluid extraction (SFE) [13]; pressurized liquid extraction (PLE) [14]; cloud point extraction [7]; matrix solid-phase dispersion (MSPD) [15]; and ultrasound-assisted dispersive solid-phase microextraction [16]. (Solvent mixtures described for sample preparation or mobile phases in this chapter are given in volume proportions, v/v, unless otherwise noted.)

Sample and standard solutions are applied manually to the plate as spots, with the help of a micropipette or syringe, or automatically, as bands or spots, with a commercial instrument. Stationary phases reported were usually 20 × 20 cm or 10 × 20 cm glass, aluminum, or plastic (polyethylene terephthalate [PET])-backed TLC or HPTLC normal-phase (NP) silica gel 60 plates (6-nm pore size) with an organic binder and with or without a fluorescent phosphor (plates with the designation F or $F_{254}$, or UV or $UV_{254}$ contain a phosphor that emits green light when irradiated with 254-nm ultraviolet (UV) radiation in order to facilitate fluorescence quenching detection of compounds that absorb this radiation as dark zones. Lichrospher HPTLC plates have layers with spherical rather than the usual irregular-shaped particles of silica gel [17]. Plates designated with a G contain gypsum inorganic binder. Cellulose, metal-cation-exchanged cellulose [18], polyamide, alumina, reversed-phase (RP) octyldecylsilyl (C18 or RP18) and octylsilyl (C8 or RP8) chemically bonded silica gel, and other chemically bonded silica gel plates such as silanized silica gel (dimethyl modified, C2 or RP2) and amino ($NH_2$), diol, and cyano (CN) [19] have been used when the necessary selectivity and efficiency for the colorant separation are not provided by plain silica gel. Plates designated with a W are wettable with water. Thicker layers are often used in preparative TLC (PTLC) for the purification of separated fractions prior to further analysis [20]. Commercial plates referred to in this chapter are manufactured by Merck Millipore, unless otherwise noted.

Homemade plates were also prepared for food dye separations. Ultra-thin-layer chromatography (UTLC) was carried out on electrospun glassy carbon nanofibers, 2 × 6 cm, with a ca. 200- to 350-nm particle diameter and 15-μm layer thickness; applicability was demonstrated by the separation of a mixture of six dyes, including rhodamine B, a widely used food dye. Plates were spotted with 50 nL of sample by using a 250-um internal diameter (id) glass capillary and developed with 2-propanol in a 200-mL sealed equilibrated cylindrical glass jar. A Spectroline digital documentation system was used for UTLC visualization [21]. Plates (3 × 3 cm) with a monolithic layer (100-μm thickness) of silica were prepared and used to separate two three-colorant mixtures (methyl red, malachite green, and bromocresol green and turmeric, erythrosine, and brilliant green) in a 2-cm run with toluene-methanol (8:2) [22].

NP silica gel UTLC plates were prepared with 4.6-μm to 5.3-μm layer thickness and several types of in-plane macropore anisotropies, using the glancing angle deposition (GLAD) approach to engineering nanostructured thin films. Performance of two new media, isotropic vertical posts and anisotropic blade-like films, were compared with that of anisotropic chevron media, using dimethyl yellow separated from a test dye mixture distances <10 mm in a horizontal chamber after the application of 25-nL spots. A new approach involving a CanoScan 5000F flatbed scanner and custom numerical image analysis software was required to extract chromatograms from angled tracks within the anisotropic media. Limit of detection (LOD) was 10+/–4 and 11+/–3 ng for the vertical posts and bladelike media, respectively. Theoretical plate heights (HETP) varied with film structure between 12 μm and 28 μm. It was concluded that the macropore anisotropies engineered by GLAD may expand the capabilities of future UTLC stationary phases [23].

Plate development with the mobile phase is usually done at laboratory temperature in a large-volume, mobile-phase vapor-saturated N-chamber in the one-dimensional (1D) ascending mode. Two-dimensional (2D) [24], manual multiple development, and automated multiple development (AMD) [25] with isocratic or gradient elution have also been reported for the increased resolution of colorants.

The method used for choosing the mobile phase is seldom given in research papers, but it is almost always based on a guided trial-and-error approach employing the analyst's personal knowledge of the analyte, layer, and mobile-phase properties and literature searching for previous systems applied for the required separation. Mobile-phase compositions given in this chapter are in v/v proportions, unless otherwise noted; "ammonia" stands for concentrated ammonium hydroxide, 28%–30% $NH_3$; "acetic acid" stands for glacial acetic acid; and "water" is deionized or distilled.

The most widely used TLC equipment and instrumentation reported in the colorant analysis literature is manufactured by CAMAG, including sample and standard solution spot or band application by the fully automated Automatic TLC Sampler 3 or 4 (ATS 3 or 4) or semiautomated Linomat 4 or 5; development in a twin trough N-chamber (TTC), Automatic Developing Chamber (ADC 2), Horizontal Development Chamber (HDC 2), or Automated Multiple Development device (AMD 2); detection reagent application with a Chromatogram Immersion Device; bioaurography with a BioLuminizer; chromatogram documentation with a DigiStore system or TLC Visualizer; slit scanning densitometry with a TLC Scanner 3 or 4 with winCATS software; videodensitometry with VideoScan software; and an elution head type TLC-MS Interface. Desaga equipment reportedly used includes an AS-30 applicator, DC-MAT N-chamber, horizontal H-Chamber, and CD-60 densitometer.

## 4.3 STUDENT EXPERIMENTS INVOLVING COLORANT ANALYSIS FOR INTRODUCTORY ORGANIC CHEMISTRY COURSES

A laboratory exercise was detailed, covering three sessions of 3–4 h each, involving continuous and discontinuous extraction, vacuum and atmospheric pressure distillation, and TLC. Samples analyzed for curcuminoids were natural products (rhizomes of turmeric) and processed food from a grocery store (dehydrated vegetable soup and chicken-flavored instant bouillon) [26].

An experiment involving the separation and isolation of the natural carotenoid food dye bixin from annatto seeds in the following steps can be completed in a 4-h laboratory session. The carotenoids are extracted from seeds with dichloromethane or acetone; the students then choose the best mobile phase out of five alternatives to separate bixin from the mixed carotenoid extract; the mixture is separated using column chromatography to obtain 20–25 mg of bixin from 2 g of crushed seeds; and the isolate is compared by UV/visible (UV-vis) and infrared (IR) spectrometry with commercial samples [27].

To illustrate the application of chemistry to everyday lives, students synthesize FD&C dyes yellow 5 and yellow 6. These are then run as standards on TLC plates along with samples of food items brought in by the students to determine the presence in the consumed products [28].

## 4.4 APPLICATION TO STUDIES OF SYNTHETIC COLORANTS IN FOODS

This and the next section contain selected applications of TLC in the study of synthetic and natural food colorants, respectively. In each, the techniques, materials, and instruments that were reported for each step of the TLC process are given. Synthetic food dyes are brightly colored and stable and are widely added as adulterant ingredients to foods and beverages for masking defects such as loss of natural pigments during processing and storage or for intensifying colors to give the impression of better products, thereby increasing consumers' appetites and their consumption of these products. TLC is an important method for their monitoring, especially if the additives have negative effects on health.

### 4.4.1 Separation, Detection, and Identification

Tartrazine, amaranth, sunset yellow, and brilliant blue were analyzed in commercial carbonated orange and grape soft drinks produced in Ceara, Brazil. Dyes were extracted using Sep-Pack C18 SPE and identified by TLC. Quantification was done by ion-pair column HPTLC with a UV diode array detector (HPLC-DAD). Standards and sample extracts were applied as spots to 20 × 20 cm silica gel plastic plates with a 5-μL capillary; the plates were developed in a glass chamber with isopropanol-ammonia (8:3); and the dyes were identified by the comparison of $R_f$ values and colors [29].

RP8 layers and mobile phases of methanol-water-acetic acid (4:5:1), methanol-acetate buffer (45:55), and methanol-water (1:1) were shown to separate allura red AC, patent blue V, and brilliant black PN food dyes. The method was proposed to analyze these dyes added to enhance the color of fruit products and fish and shellfish paste in the food industry [30].

Twelve water-based felt tip pens were investigated by a multi-analytical study, including TLC. Numerous synthetic food-coloring agents and pigments were identified in the pen inks, including acid yellow 23, acid red 18, acid blue 9, and pigment blue 15. Analytical TLC was carried out on samples deposited from methanol solution by using Macherey-Nagel ALUGRAM Xtra 0.2-mm-layer-thickness Sil G/$UV_{254}$ precoated aluminum plates with an acidic mobile phase, *n*-butanol-water-acetic acid (60:24:16), and a basic mobile phase, *n*-butanol-ethanol-ammonia (5:2:3). PTLC on Analtech (now Miles Scientific) Uniplate SIL GF/$UV_{254}$ 1.5-mm-thick layers was used to purify samples on a few tens of milligram scale prior to further analysis [31].

Different types of fruit juices (e.g., cherry and apple) and wines were investigated in terms of their synthetic and natural food-coloring constituents. Samples were acidified with acetic acid, heated, and then subjected to SPE with an alumina cartridge and aqueous ammonia solution (25 mL/100 mL). Analysis was on a 10 × 10 cm Silufol plate developed with pyridine-isoamyl alcohol-isobutanol-ethanol-25 mg/100 mL aqueous ammonia solution (3:3:3:4:4). Synthetic food colors were identified in the presence of natural colorants based on TLC analysis ($R_f$ values compared to standard dyes), color characteristic measurements, and IR spectra [32].

The TLC retention behavior of 18 food-colorant standards was tabulated under thermostatic conditions (30°C) for silica gel W F and cellulose microplates (5 × 5 cm) and mobile phases comprising 100% water, methanol, and dichloromethane; 20%–80% methanol-water; and 10%–80% methanol-dichloromethane to facilitate the selection of separation conditions. Spot patterns on developed plates were acquired under visible light by using a Plustek OpticPro S28 USB scanner and Image Folio v. 4.50.03 software, and quantitative data (spot positions, peak profiles, and peak geometry) were acquired using ImageJ software [33].

Pararosaniline, auramine O, and rhodamine B were identified in chili sauce, curry paste, soup powder, shrimp powder, gochujang, and tandoori chicken by TLC on RP18 plates. C18 SPE was used for sample preparation, and standard and extract solutions were applied with glass capillary tubes. The mobile phases were 2-butanone-methanol-5% $Na_2SO_4$ (1:1:1) and 2-butanone-methanol-1.6 M ammonium formate (pH 2.5) (7:2:7). Developed plates were observed under white and 254- and 366-nm UV light and documented using a TLC Visualizer. C18 column gradient elution HPLC was used for quantification [34].

Orange dye, ponceau 4R, sunset yellow, and tartrazine were identified in energy drinks, juices, and soft drinks by extraction with 5% acetic acid solution onto wool thread, followed by removal from the thread by soaking and boiling with 1 M ammonia solution (AOAC 2005 standard method) and silica gel TLC using propanol-ammonia (60:15). Sample components were identified by comparing $R_f$ values and colors of samples and standards [35].

The separation of brilliant blue, tartrazine, and carmoisine was improved on polyaniline-modified silica gel compared with conventional silica gel, with the use of methanol-water (1:9) mobile phase. The layer material was prepared by in situ oxidative polymerization of aniline in the presence of different amounts of silica gel by using potassium persulfate as the oxidizing agent [36].

Pigments used in packaging can migrate into foods and compromise quality and safety. A nine-step gradient elution HPTLC-AMD method based on ethyl acetate, methanol, and water was devised for QC of 124 different pigment formulations used in packaging, including color components, additives, and coating materials, in very complex, matrix-rich samples. Hyphenation with mass, NMR, and FTIR spectrometries allowed the assignment of single unknown pigment components and with *Aliivibrio fischeri* (formerly *Vibrio fischeri*) bioassay revealed the biological potential of different pigment compounds [25].

Rather than the AMD commercial device, stepwise gradient elution separation of a 19-component test dye mixture was carried out on RP18 W plates in a nonautomatic instrument, based on the Chromdes horizontal Teflon chamber [37]. The operation procedure described is complicated, and the need to construct a homemade instrument is probably not practical for most analysts.

A polyamide TLC limit test method was devised; it showed that LODs for ponceau 4R, amaranth, tartrazine, sunset yellow, and brilliant blue ranged from 2.06 to 10.1 ng. Beverages were applied directly, and gelatinous and solid food samples were applied after ultrasonic extraction with water-ammonia (99:1). Applied acidified sample solutions were cleaned up by on-plate SPE with 95% ethanol-water-acetic acid (5:5:0.1) and then developed in a TTC with 95% ethanol-water-ammonia (75:14:1). The chromatograms were viewed in the back-light mode with a light-emitting diode (LED) array light [38].

### 4.4.2 Thin-Layer Chromatography-Raman Spectrometry

Surface-enhanced Raman spectrometry (SERS) is combined with TLC to improve the spectrometric sensitivity by eliminating interferences from a mixture [39]. The interference of the mobile-phase background is also eliminated, because the plate is usually dried before the detection. Examples for food dyes are given in this section.

The mutagenic and carcinogenic azo dye Sudan I, used as a common adulterant in spices to impart a red color, was determined in paprika by using extraction by sonication with acetonitrile, TLC of the extract on a fabricated molecularly imprinted polymers (MIPs)-TLC plate with hexane-chloroform (1:1) mobile phase, and SERS detection of the dye selectively fixed at the origin while interferences migrated up the plate with both portable and benchtop spectrometers. A gold colloid SERS substrate enhanced the Raman intensity for Sudan I in the MIP-TLC system. Principal components analysis (PCA) plot and partial least squares (PLS) regression chemometric models were applied for quantification against 5–100 ppm standards [40].

Semiquantitative TLC-SERS was described with calibration curves pre-built in the laboratory and then single-point calibration used in the field for onsite food QC inspection. The method was tested for the analysis of methanol chili oil extract of rhodamine B by TLC on silica gel 60 F plates, developed with ethyl acetate-ethanol (2:1). The dye spot was scratched off, dissolved in 30% aqueous acetic acid, mixed with melamine internal standard, and deposited onto aluminum foil containing silver nanoparticles. Removal from the plate eliminated the layer background in the SERS spectra to give an LOD of $1.00 \times 10^{-7}$ M, and recovery was 66.1%–110% [41].

Highly specific SERS analysis of illegally used colorant Sudan I in chili powder, sauce, and oil and food oil without sample pretreatment was performed on fabricated TLC plates from diatomite as the stationary phase. Liquid samples (method of sample preparation was not described) were spotted on the fabricated diatomite earth TLC plates and developed with cyclohexane-ethyl acetate (6:1). After separation, prepared gold nanoparticles were deposited, and SERS spectra were acquired using a Horiba Jobin Yvon Lab Ram HR800 Raman microscope equipped with a charged coupled device (CCD) camera. Intensity was improved more than 10 times compared with a commercial silica gel TLC plate, with detection down to 1 ppm or 0.5 ng/spot [42].

Rhodamine 6G was used as a probe to evaluate a fabricated TLC plate with metal-organic-framework-modified gold nanoparticles as a substrate on which all separated compounds, including those that are overlapping and invisible, could be detected by a point-by-point SERS scan along the development direction. Samples were applied to the plates by using a quantitative capillary, and after drying, the plate was developed inside a mobile-phase saturated traditional chamber. Then, the mobile phase was volatilized, the plate was wetted by water to protect it from laser damage, and SERS spectra were collected by using a B&W Tek BWS415 portable Raman spectrometer. The detection method was sensitive and rapid, and the new substrate exhibited a long shelf life [43].

### 4.4.3 Thin-Layer Chromatography-Mass Spectrometry

Additional reports of TLC hyphenated with MS in food dye analysis are cited in other sections of this chapter.

A TLC/LIAD (laser-induced acoustic desorption)/ESI (electrospray ionization)-ion trap (Bruker Daltonics esquire 3000 + mass analyzer) MS method allowed components in mixtures of dye standards (FD&C Green No. 3, FD&C Red No. 3, and eriochromcyanin) and rosemary essential oil to be separated and rapidly characterized. LIAD analysis was performed by radiating the rear side of an aluminum-backed C18 Fs (s designates an acid stable indicator) or silica gel TLC plate with a Q-switched Nd:YAG pulsed IR laser. Acoustic and shock waves were efficiently generated and transferred to ablate the separated analytes from the layer by attaching a glass slide to the rear of the plate with a gap filled by glycerol [44].

Electrospray-assisted laser desorption/ionization (ELDI) MS and tandem MS (MS/MS) were used under ambient conditions to characterize a mixture of food dyes (FD&C red No. 3, blue No. 1, and green No. 3), separated at the central track of a 3 × 0.5 cm) C18 TLC plate. The plate was placed on an acrylic sample holder set in front of the sample skimmer of an ion trap mass analyzer, and the dyes were desorbed by pushing the sample holder into the path of pulsed nitrogen laser beam with a syringe pump into an ESI plume, where they were ionized through reaction with charged species generated by electrospraying a water-methanol solution. The LOD of the red dye was $10^{-6}$ M, and its calibration curve was linear from $10^{-3}$ to $10^{-6}$ M [45].

Quantitative performance data were reported for the elution mode TLC-MS Interface that hyphenates TLC to single quadrupole ESI MS [46]. Analysis of commercial lipophilic dye mixtures on silica gel 60 with toluene mobile phase using an ATS 4, ADC 2 with humidity control, and a TTC showed mean precisions ($n = 5$) below 10% and mean calibration correlation of 0.9975. HPTLC-MS enabled the identification of incorrectly assigned unknown dyes in mixtures.

A plasma-assisted multiwavelength (1064, 532, and 355 nm) laser desorption ionization MS (PAMLDI-MS) system successfully integrated TLC, multiwavelength laser ablation, and the excited-state plasma from direct analysis in real time (DART)-MS to effectively separate and selectively identify a model mixture of rhodamine B, fluorescein, and Sudan III, with an LOD of 5 ng/mm$^2$. Use of this laser provided broad possibilities for wavelength-dependent desorption of different compound types. The silica gel TLC-MS was carried out using manual sample and standard application with a capillary tube, dichloromethane-ethanol-ammonia (66.1:33.1:0.8) mobile phase, a Panasonic FP1 digital camera for plate photography, and an Agilent XCT ion trap mass spectrometer with a DART ion source [47].

### 4.4.4 Quantitative Analysis

A combination of image processing of scanned chromatograms with HPTLC on Macherey-Nagel Nano-SIL amino-modified plates and isopropanol-diethyl ether-ammonia (2:2:1) mobile phase was used in the quantitative determination of sunset yellow, azorubine, and tartrazine in dried, concentrated juice products dissolved in water, with no additional sample preparation. Samples and standards were applied with a microsyringe, ascending development was in a saturated chamber, and chromatograms were scanned using the TrueColor setting of an HP ScanJet 3970 and digitally processed by Macherey-Nagel TLC Scanner software. Linearity correlation coefficients ranged from 0.995 to 0.998, LOD from 5.21 to 9.34 ng/spot, limit of quantification (LOQ) from 10.2 to 18.1 ng/spot, and recoveries at two concentration levels from 96.4% to 103% [48].

A rapid HPTLC method was developed for screening large numbers of samples of spices (paprika, chili, and curry powder) for the presence of illegal dyes. Extraction was carried out by turbomixing with acetonitrile, and extracts were cleaned up through a Chromabond silica gel SPE cartridge. HPTLC was on an RP18 layer with acetonitrile-ammonia (95:5) mobile phase in an unsaturated TTC after sample application with a Linomat 5, and the plates were documented with a DigiStore and evaluated by densitometry with a TLC Scanner 3. The method was validated regarding precision and recovery (accuracy), and it was being routinely used by the Official Food Control Authority of the Canton of Zurich since 2007 [49].

One of the most important studies published illustrates the advantages of TLC for the identification and quantification of 25 water-soluble dyes in bakery inks and energy drinks directly applied after dilution [50]. Horizontal development of 20 × 10 cm silica gel 60 F HPTLC plates in an HDC 2 chamber took 12 min for 40 runs in parallel, using 8 mL of ethyl acetate-methanol-water-acetic acid (65:23:11:1) mobile phase, up to a migration distance of 50 mm. The total analysis time was inclusive of application with an ATS 4, and migration took 60 min for 40 runs, which translated to 1.5 min/run with a solvent consumption of 200 μL. The analysis throughput was tripled to 1000 runs/8 h day by switching between application, development, and evaluation stations in a 20-min interval. Both multiwave classical absorption densitometry with a TLC Scanner 3 and digital evaluation of plate images with VideoScan software were used for fully validated quantification. If necessary for confirmation, online ESI-mass spectra were recorded in 1 min by using a ChromXtract interface, forerunner to the CAMAG TLC-MS Interface. Figure 4.1 shows analog curves resulting from the transformation of 12-bit CCD camera images with the VideoScan software, and Figure 4.2 illustrates quantification by densitometry performed with 4-point calibration via peak height or area.

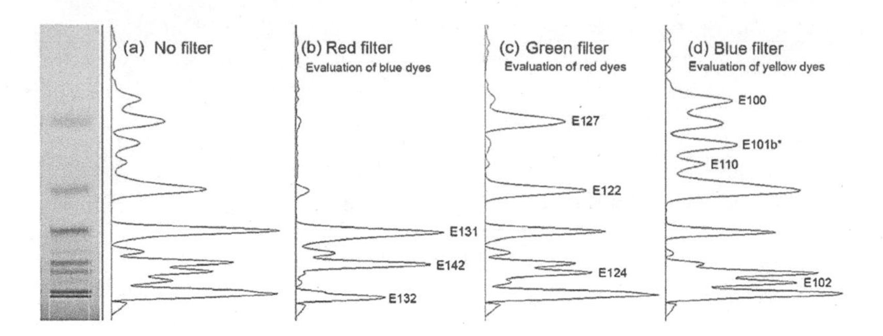


**FIGURE 4.1** Digital evaluation of a track (mixture of E131 [patent blue V], E142 [acid green 50], E132 [indigoline], E127 [erythrosine], 122 [azorubine], E124 [ponceau 4R], E100 [curcumin], E101b* [riboflavin 5-phosphate], E110 [sunset yellow FCF], and E102 [tartrazine]) transformed into an analog curve chromatogram (a) and using different filters (b–d) to separate the colors. (Reproduced from Morlock, G.E. and Oellig, C., *J. AOAC Int.*, 92, 745–756, 2009. With permission of AOAC International.)

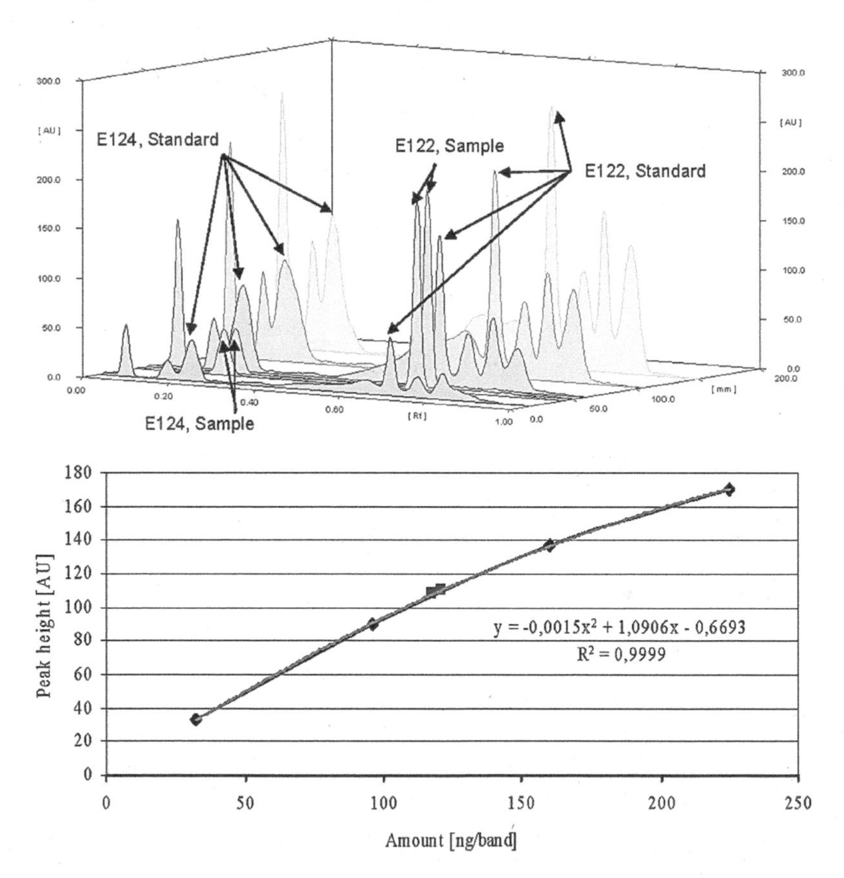


**FIGURE 4.2** Absorbance scans (analog curves) for the determination of E122 and E124 in bakery ink formulation and the calibration function of E124. (Reproduced from Morlock, G.E. and Oellig, C., *J. AOAC Int.*, 92, 745–756, 2009. With permission of AOAC International.)

A TLC method developed by Tuzimski [51] for the identification and quantitative analysis of drinks and drops for tartrazine, quinolone yellow, azorubine, ponceay 4R, allura red AC, patent blue V, and brilliant blue FCF comprised C18 SPE, sample application with a Linomat 5, RP18 W F or CN F layers developed in a Chromdes horizontal Teflon DS chamber with methanol-acetate or citric buffer containing diethylamine or octane-1-sulfonic acid sodium salt, and densitometry in the range of 200–800 nm with a J&M DAD scanner. LOD values ranged from 32 to 119 ng/zone, LOQ from 95 to 362 ng/zone, and correlation coefficients from 0.990 to 0.998. An earlier study by Tuzimski and Wozniak [19] employed the same SPE, TLC on RP 18 W layers, and DAD densitometry techniques to identify and quantify tartrazine and sunset yellow in sparkling fruit-flavored beverages, with the respective mobile phases methanol-pH 5.6 acetate buffer (3:7) and tetrahydrofuran-pH 3.5 acetate buffer (7:13).

Food colorants sunset yellow and amaranth were determined from pharmaceutical powders by UAE with methanol-ammonia (9:1), purification by ion-pair SPE on C18 cartridges by using hexadecyltrimethylammonium bromide (CTAB), standard and sample application with a Linomat 5, silica gel G TLC by using isopropanol-ammonia-0.01 M CTAB (7:3:2) mobile phase, and quantification with a Shimadzu CS-9000 dual-wavelength flying-spot densitometer at 485 nm and 520 nm, respectively. LOQ values were 0.29 and 0.46 ug/mL, and recoveries were 99.3+/−3.73.7 and 98.8+/−1.2%, respectively [52].

The analysis of tartrazine from mustard was carried out on 20 × 10 cm silica gel G plates developed with isopropanol-ammonia (7:3) in a normal chamber (Desaga). The dye was extracted from various mustards purchased from a local market by UAE with methanol-ammonia (9:1); further purified by C18 SPE, with CTAB as an ion-pair reagent; and applied to plates with a Linomat 5. Standard addition recoveries from spiked samples were around 98%, with relative standard deviation (RSD) of 17.4%. Results from densitometry with a Shimadazu 9000 at 425 nm were compared with those obtained by using digital processing of plates scanned with a BioDit TLC Scanner, using ImageDecipher TLC software [53]. The same research group described the determination of azorubin and sunset yellow in jelly dessert powders, using silica gel 60 plates, *n*-butanol-acetic acid ethanol-water (10:2:1:5) mobile phase, and the same instrumentation. Recoveries were 96.9% and 95.7%, LOD 9 and 18 ng/zone, LOQ 11 and 23 ng/zone, respectively, and RSD values for both were less than 2% [54].

Synthetic colorants tartrazine, quinolone yellow, sunset yellow FCF, azorubin, amaranth, ponceau 4R, allura red AC, patent blue V, indigo carmine, and brilliant blue FCF were quantified by HPTLC on RP18 plates with 0.5 M ammonium sulfate in ethanol-water (3:7) mobile phase. $R_f$ values were in the range of 0.17–0.64; linearity was in the range of 20–300 ng/zone; LOD and LOQ were 2 and 3 ng/zone, respectively; recovery by standard addition was 81–108%; and intermediate intraday and interday precision was less than 5% (n = 3). A Linomat 5, saturated horizontal chamber, and Scanner 3 at 549, 500, and 625 nm were utilized [55].

The Office Chromatography concept has been introduced by combining office technologies, miniaturized TLC, and printing of materials. In a 2015 study [56], precise printing was carried out from a modified bubble-jet printer of a food dye mixture's (tartrazine, azophloxin, crystal ponceau, lissamine green, and patent blue V) sample solutions on a 1.5 × 6.0 cm carbon nanotube templated (CNT)-UTLC plates, obtained from the chemistry department of Brigham Young University [57], that were developed in a custom-built glass chamber having 12 × 40 mm base area and 70-mm height, with ethanol-methanol-water- acetic acid (150:35:25:1), followed by densitometric evaluation, digital documentation with a CanoScan 9000F, image evaluation with VideoScan evaluation software, and DART ion-trap MS scanning. The accuracy of printing was proven by quantitative validation data. In an earlier paper [58], the Morlock research group explored $SiO_2$, $Al_2O_3$, $TiO_2$, and $ZrO_2$ nanostructured GLAD UTLC media for the separation of carotenoids and synthetic food dyes applied with an inkjet printer in the Office Chromatography concept. In a later paper [59], an electrospun nanofibrous stationary phase with fiber diameters of 150–225 nm and a thickness of 25 µm was manufactured, and mixtures of water-soluble food dyes were printed on it by using a modified Canon Pixma iP 3000 bubble-jet printer to illustrate its capabilities for UTLC in a homemade microchamber (4.5 × 1.5 × 5.5 cm), using methanol-toluene-ammonia (40:57:3) mobile phase, videodensitometry with a CanoScan 9000F flatbed scanner and VideoScan software, and MS coupled through the TLC-MS Interface.

A densitometric method given for the determination of tartrazine, ponceau 4R, brilliant blue, orange yellow, and azorubine in mouthwashes and beverages comprised SPE on $NH_2$-bonded silica gel, with 0.01 M aqueous sodium hydroxide as the eluent and silica gel 60 TLC (0.25 mm layer thickness, 10 × 20 cm plates), with chloroform-isopropanol-ammonia (1:3:1) mobile phase [60]. Standard and sample solutions were applied with an AS-30, and reflectance densitometry was with a CD-60 at appropriate maximum absorbance wavelengths. Recoveries of all dyes were more than 90%, and repeatability RSD was 4.1% or less.

The most recently published quantitative method was the rapid and inexpensive videodensitometry determination of Sudan red dyes in spices and spice mixtures. Chili paste was extracted with acetonitrile; band-wise application was done manually with a 5-µL glass pipette in small dots; RP18 F plates were developed vertically in a saturated chamber with acetonitrile-methanol-25% aqueous ammonia solution (8:1.8: 0.2); and quantification was with a 16-bit Plustek OpticSlim 500 flatbed scanner and Presto!ImageFolio 4 acquisition software and ImageTLC in PureBasic evaluation program. LOD was 500 ppb, and range of linearity was 20–500 ng [61].

### 4.4.5 Degradation Studies

Azo dyes are found in foods after authorized and unauthorized use. The greater toxicity of degradation products compared with the parent compound may be the cause of prohibition of use of some dyes. The azo dye methyl red was used as a model food dye in an HPTLC study of degradation by oxidation (large effect found) and sonication (negligible effect found). Silica gel 60 F glass plates were spotted with an ATS 4; development with chloroform was in a TTC (sharp, well-defined peak at $R_f$ 0.10); documentation was with a DigiStore 2 system, with illuminator Reprostar 3 and Baumer optronic DXA252 digital camera in the reflectance mode at 254 nm; and quantitative densitometry was with a TLC Scanner 3 at the 501-nm absorption maximum [62].

The decolorization of acid blue 158 under microaerophilic condition in wastewater potentially from a food industry site was shown to be mainly a function of concentration and temperature when monitored by TLC and FTIR spectrometry. Silica gel plates were spotted with glass capillaries and developed with pyridine-methanol-ammonia-water (5:6:1:16) mobile phase [63].

TLC, MS, and FTIR spectrometry were applied to study the degradation of methyl red as a model azo dye by *Saccharomyces cerevisiae* ATCC 9763 in food industry effluents. Silica gel 60 F plates were developed with methanol-water (9:1), and individual spots were visualized by UV irradiation at 312 nm. The two confirmed degradation products were N,N-dimethyl-p-phenylene diamine and 2-aminobenzoic acid [64].

### 4.4.6 Monitoring of Biosynthesis

Chemical production of indigo dye used in the textile, food, and pharmaceutical industries was replaced by a highly productive environmentally friendly biosynthesis alternative, using an indigenously isolated naphthalene-degrading strain *Pseudomonas* sp. HOB1 producing the blue pigment when indole was added in the growth medium. The pigment was analyzed by extraction from the whole culture broth by shaking with chloroform, application of the extract and commercial grade standard with a Linomat 5 onto silica gel 60 F HPTLC plates, and development with chloroform-methanol (15:1). Visual detection of the colored zones was done [65].

A study focused on the investigation of the effect of the antifungal agent fluconazole on red pigment production from *Monascus purpureus* (NMCC-PF01) confirmed the presence of rubropuctamin, based on TLC, visible spectrometry, and MS. The red pigment was extracted from fungicide-treated fermentation medium by ethanol, and the pigment fraction was obtained by gradient elution silica gel column chromatography of the extract. Further purification was performed on a silica gel 60 plate with the mobile phase chloroform-methanol-acetic acid (285:21:9). The dominant red band was scraped from the plate, redissolved in ethanol, and characterized at 500 nm by using a Shimadzu UVmini-1240 spectrometer and a Waters ESI Q-TOF micromass spectrometer. Production of citrinin-free pigments by the fermentation process was also confirmed [66].

## 4.5 APPLICATION TO STUDIES OF NATURAL COLORANTS IN FOODS

Natural colorants extracted from plants, animals, or microorganisms have more limited application in food industries because of disadvantages of some, owing to high cost, poor coloring ability, and ease of discoloration compared with synthetic dyes [67]. However, TLC is often used to analyze these colorants added to or naturally contained in foods, as described in the following subsections.

### 4.5.1 Anthocyanins

Anthocyanins, which are in the class of flavonoids, are glycosides of anthocyanidins. The term *anthocyanes* includes both anthocyanins and anthocyanidins.

TLC of the pigments (mostly anthocyanins) from different kinds of red wines (Cabernet Sauvignon, Merlot, and Burgundy) on RP18 plates developed with acetonitrile-water-formic acid (20:29:1) in a saturated N-chamber was used to identify the origin and to detect adulteration. Samples were applied with a microsyringe. Evaluation was made in the visible light and under 366-nm UV light and by spraying with 0.5 mg/mL methanolic DPPH* (2,2-diphenyl-1-picrylhydrazyl radical) solution [68].

The major anthocyanin colorant isolated from ogaja fruit was identified by HPLC-MS to be cyanidin-3-O-sambubioside. After acid hydrolysis of the anthocyanin fraction purified by recycling preparative HPLC, it was structurally confirmed by TLC on a 10 × 20 cm Si 50,000 HPTLC plate that the sugar moieties linked to cyanidin were glucose and xylose. Ascending development was in a chamber containing *n*-propanol-water-triethylamine-ammonia (80:20:0.2:4), and monosaccharides were detected as colored zones by spraying with 0.3% N-(1-naphthyl)-ethylenediamine and 5% sulfuric acid in methanol and by heating at 121°C for 10 min [69].

TLC profiles of anthocyanins and anthocyanidins present in eight berry fruits were used as reference fingerprints to evaluate the authenticity and conformity to the label information of fruit juices from the market. Separations of sample extracts were performed on Macherey-Nagel CEL 300-25 cellulose glass plates with HCl-acetic acid-water (40:4:12) mobile phase in a saturated normal chamber, and chromatogram scanning was done with a CD-60 densitometry at 520 nm in the reflectance-absorption mode [70].

An HPTLC method was developed to analyze the anthocyanin patterns of colored wheat varieties. Sample preparation was simplified to extraction and filtration, exploiting the matrix accommodation of TLC. Anthocyanin patterns were used to verify scoring of the grain color. Assignment to grain color classes was done by both visual inspection of chromatograms and multivariate analysis of densitometric data [71]. An earlier HPTLC study of colored wheat extracts was for combined grains of blue aleurone and purple pericarp genotypes. The best separation was achieved on $NH_2$ layers with ethyl acetate-2-butanone-water-formic acid (14:6:3:3) mobile phase. HPTLC-UV/Vis fingerprints allowed clear differentiation of the genotypes. HPTLC-ESI/MS allowed further characterization and differentiation of anthocyanin profiles [72].

The ancient blue/black cultivar “Millo Corvo” was characterized by TLC and HPLC, and it was demonstrated that it accumulates high amounts of anthocyanins consisting mainly of cyanidin. Pericarp powder and aleurone layers of kernals were prepared by boiling with 2 M HCl and extracted with isoamyl alcohol. Anthocyanin standards and extracts were loaded on precoated Macherey-Nagel Polygram cellulose plastic plates (Cel 300) and developed with formic acid-HCl-water (5:2:3) mobile phase. The plates were photographed with a Canon A430 digital camera, using both white and UV illumination [73]. The same TLC procedures were used to show that the anthocyanin content in the functional food colored polenta was reduced by 22% by cooking, but the remaining anthocyanins exhibited 2-folds higher antioxidant capacity (DPPH* assay) [74].

A simple HPTLC method was given to allow high-throughput profiling of phenolic compounds of microwaved roots from 295 sweet potato varieties and breeding lines, quantify the content of anthocyanins and caffeoylquinic acid derivatives, and determine the respective contributions to the antioxidant activity of sweet potato methanolic extracts, using the DPPH* test. Analyses were performed on 20 × 10 cm silica gel 60 F plates with a CAMAG system, including an ATS 4, ADC 2, TLC Visualizer, and TLC Scanner 4. The mobile phase was ethyl acetate-methanol-acetic acid-formic acid-water (27:2:2:2:2); digital images of plates were documented using the Visualizer with 12 bit CCD digital camera; and plates were scanned at 330 nm for phenolic acids and in white light for anthocyanins [75].

TLC detection was carried out for two anthocyanins (cyanin and keracyanin) and two anthocyanidins (pelargonidin and delphinidin) in a selection of commercial and homemade fruit juices, and antioxidant activity of plant samples was determined by DPPH* and ABTS (2,2′-azino-di-[3-ethylbenzthiazoline] sulfonate) dot blot and spectrometry tests. Relation between the occurrence of plant pigments in the investigated fruit preparations and their antioxidant properties was established. Juices and dried plants were prepared as methanolic solutions and applied onto microcrystalline cellulose layers, which were developed in saturated standard chromatographic chambers with concentrated HCl-80% formic acid-water (9:46:90) for pelargonidin determination and 80% formic acid-water-*n*-butanol (16:19:65) for the other three analytes. Pigment zones were detected by scanning with a CD-60 densitometer in visible and 254-nm UV light and after visualization with ammonia vapors. Pigments were identified by comparison of sample and standard zone $R_f$ values and by TLC-MS, using a CAMAG interface and Varian 500-MS ESI spectrometer [76].

A quantitative HPTLC method was developed for 11 anthocyanins in various foods and beverages. Extracts were obtained by stirring with 0.5% HCl in methanol; samples and standards were applied with an ATS 4; the best separations were achieved on silica gel 60 F plates with ethyl acetate-2-butanone-formic acid-water (7:3:1.2:0.8) mobile phase in an ADC 2 for anthocyanins and ethyl acetate-toluene-formic acid-water (10:3:0.8:1.2) for anthocyanidins. Absorbance measurement with a TLC Scanner 3 was performed using multiwavelength scan at 505 or 510, 520, 530, and 555 nm. The method was validated for linearity, LOQ, precision, accuracy, and ruggedness. Unknown anthocyanins were analyzed by HPTLC-ESI/MS, and EDA with *Aliivibrio fischeri* bacteria (BioLuminizer) was used to demonstrate radical scavenging properties [77].

Less expensive cellulose TLC plates were shown to outperform silica gel HPTLC plates for the authentication of 18 selected juices, syrups, nectars, and noncarbonated drinks and targeting anthocyanes as authenticity markers. Chlorides of cyanin, keracyanin, pelargonidin, and delphinidin were chosen as external standards for the development of calibration curves with a 60-CD densitometer and Pro-Quant software, and LOD levels were one order of magnitude lower for TLC. Samples diluted with methanol and HCl were chromatographed in a saturated standard flat-bottom chamber on the cellulose TLC plates acetic acid-water-*n*-butanol (16:19:65). Separated bands were confirmed with the TLC-MS Interface and a Varian 500-MS mass spectrometer [78].

### 4.5.2 Carotenoids

SFE was used to extract bixin from annatto seeds, prior to the profiling of extracts on Xtra SIL G plates (Macherey-Nagel) with chloroform-ethanol-acetic (95:5:1) mobile phase. SFE satisfies the requirements of a green extraction method being emphasized in the development of analyses today [13].

The carotenoids β-carotene, lycopene, lutein, astaxanthin, and zeaxanthin were separated and identified in vegetable products and a dietary supplement by TLC on silica gel 10 × 10 cm silica gel F plates. Extracts were applied with a 10-μL microsyringe; plates were developed with benzene-petroleum ether (9:1 and 1:9) or methanol-benzene-ethyl acetate (5:75:20); and visualization was performed under 254- and 366-nm UV light [79].

Fucoxanthin in marine algae, with potential for use as a colorant in dietary supplements and the food industry, was quantified by HPTLC and assessed for its antioxidant activity by derivatization with DPPH*. Soxhlet or shake ethanol extracts were sprayed as bands by using a Linomat 5 onto silica gel 60 F plates, which were developed in an AMD 2 using the mobile phase *n*-hexane-ethyl acetate-acetic acid (20:10:1). Chromatogram images were recorded under 366-nm UV light and white light, and quantification was performed with VideoScan Digital Image Evaluation software. PCA of HPTLC fingerprints allowed the classification of algae species into five groups, according to their chemical/antioxidant profiles [80].

Carotenoids were extracted from water spinach by using methanol-acetone-petroleum ether (1:1:1), purified by PTLC on Qingdao Haiyang Co. silica gel 60 G with mobile phase *n*-hexane-ethyl acetate-acetone-methanol (27:4:2:2) and were identified as lutein, violaxanthin, and β-carotene by comparison of $R_f$ values against standards by analytical TLC with the same system. [81].

TLC methods were reported for carotenoid, quinone, flavonoid, and anthocyanin colorants in hundreds of foods, using RP18 plates and scanning densitometry. Respective mobile phases were acetonitrile-acetone-*n*-hexane (11:7:2) and acetone-water (9:1), methanol-0.5 M oxalic acid (11:9), 2-butanone-methanol-5% sodium sulfate-5% acetic acid (3:2:5:5), and acetonitrile-0.2 M trifluoroacetic acid (1:2). Identification was made by recording visible absorption spectra [1].

Some workers have found degradation of carotenoid analytes during sample preparation and during and after TLC. For this reason, the layer used is bonded RP18 silica gel, as above [1], and operations are carried out quickly in subdued light. Another remedy for degradation was suggested in the determination of lutein and other major carotenoids in six brands of food supplements [82]. Extraction was with ethyl acetate containing 0.1% of the antioxidant BHT (2,6-di-*tert*-butyl-4-methylphenol), and TLC of samples and standards applied with an ATS 4 was on 20 × 10 cm RP18 plates developed with methanol-acetone (1:1) + 1% antioxidant TBHQ (2-*tert*-butylhydroquinone) in a TTC. Lutein, β-carotene, and lycopene were well separated; validated quantitative densitometry of lutein was carried out at 450 nm with a TLC Scanner 3; and spectra were recorded from 400 to 600 nm with the tungsten light source.

Successful quantification of food colorant β-carotene on the adsorbent layer neutral aluminum oxide 60 F was completed in dietary supplement capsules and tablets and fruit juices [83]. The mobile phase was chloroform-ethanol-acetone-ammonia (10:22:53:0.2) in a saturated TTC, and absorbance mode densitometry with a TLC Scanner 3 was at 450 nm. Extracts were prepared by shaking with ethyl acetate for supplements and chloroform for juices and applied with a Linomat 5. The method was validated for specificity, linearity, LOD, LOQ, precision, accuracy, robustness, and ruggedness per International Conference on Harmonization (ICH) 2000 guidelines.

Quantification of astaxanthin (European Union [EU]-permitted good colorant E161J) was obtained in salmon, based on chemiluminescence induced by the reaction of bis(2,4,6-triphenyl)oxidation with hydrogen peroxide and by visible light absorption after the separation of ethyl acetate-sodium hydroxide extracts applied with an ATS 4 onto silica gel 60 F plates that were developed with cyclohexane-acetone (10:2.4) mobile phase. A 16-bit CCD camera was used to measure absorption in visible light and light emission as chemiluminescence, and evaluations were made with program Image TLC (ver. 3.00), written in PureBasic (ver. 4.50). LOD and LOQ were 64 and 92 ng/band for visible absorption and 90 and 115 ng/band for chemiluminescence, respectively. The two independent measurement modes were used on a single plate to increase quantification accuracy [84].

Thermal degradation of β-carotene in edible sunflower oil (1–5 h at 100°C) was shown to be efficiently assessed by dissolving in N,N′-dimethylformamide (DMF), extraction in a separatory funnel with hexane, application of samples with an ATS 3, HPTLC on silica gel 60 F aluminum plates with petroleum ether-hexane-acetone (2:3:1) as the mobile phase in a TTC, and densitometric quantification with a TLC Scanner 3 at 450 nm. Response was linear in the 100–600 ng range, LOD was 0.11 ng, and LOQ was 0.37 ng [85]. The same authors used HPTLC, as performed previously, to confirm the loss of β-carotene in corn, rapeseed, and sunflower oils on oxidative heating (air flow 20 L/h, 110°C, 14 h). Findings that rapeseed oil was most stable in terms of formation of polar compounds and sunflower oil least prone to oxidation were confirmed by HPLC-DAD-MS [86].

Saffron, the dried red stigmata of *Crocus sativus* L. flowers and one of the highest-priced spices used in the food industry for its color owing to crocin and crocetin content, was successfully analyzed by TLC in the 1990s [87]. More recently, TLC with image analysis has been used to obtain fingerprints for the evaluation of saffron. In a study of saffron discrimination from different origins [88], dried and ground stigmas were extracted with ethanol-water (8:2) by vortexing and ultrasonication and separated by TLC on silica gel G F plates by using horizontal development with *n*-butanol-acetic acid-water (4:1:1). Chromatograms were photographed with a Canon digital camera, model Isus 115 Hs, and intensity profiles of RGB (red, green, and blue) characteristics were produced and processed by specially designed image analysis software based on MATLAB version 7. The method allowed the comparison of different types of saffron from Iran and determination of various adulteration colorants. In a second study from Iran [89], a processing strategy was described to obtain chemical fingerprints of saffron TLC images, in order to evaluate and classify products from different areas. Stigmas were extracted by ultrasonication with methanol-acetonitrile (38:62); silica gel 60 F plates were developed with the same mobile phase; and images were captured with a Samsung galaxy 56 smartphone built-in camera in visible and UV light and imported to MATLAB environment version 9.3 R2014a for data analysis. Processing comprised image compression or size reduction, image pre-processing to improve quality by removing chromatographic artifacts, and transformation of image information into mathematical data.

### 4.5.3 Chlorophylls

Chlorophyll content in barley juice extract food supplements from different continents showed only slight differences based on HPTLC fingerprint profiles. Ultrasonic extraction was carried out with acetone-water (9:1), and HPTLC of extracts was performed on CAMAG equipment consisting of a Linomat 5, ADC 2, TLC visualizer, and winCATS software for data evaluation. Silica gel 60 F 10 × 10 cm plates were developed with four different mobile phases, for a distance of 6 cm; developed plates were immersed in sulfuric acid-methanol (1:9) reagent and heated at 100°C for chromatogram visualization; and digital pictures were captured at 366 nm. The mobile phases were toluene-ethyl acetate (9:1), chloroform-ethyl acetate-formic acid (2:1:1), cyclohexane-ethyl acetate (1:1), and chloroform-acetone (6:1) [90].

### 4.5.4 Curcumin

Methanol extracts of the natural colorant curcumin (1) and synthetic dyes metanil yellow (2), Sudan I (3), and Sudan IV (4) were quantified in turmeric, chili, and various mixed curry powder formulations on silica gel 60°F aluminum plates, using applicator AS-30, chamber DC-MAT, and densitometer CD-60. The mobile phases were chloroform-methanol (9:1) for compounds 1 and 2 and toluene-hexane-acetic acid (50:50:1) for compounds 3 and 4, and scanning wavelengths were 420 nm for compounds 1 and 2, 491 nm for compound 3, and 520 nm for compound 4. LOD values ranged from 7 to 43 ng/spot and LOQ from 21 to 130 ng/g for the four compounds. Standard addition recoveries (%) +/−RSD (%) were 86.7–98.7+/−1.26–3.32 for compound 1, 97.1–103+/−2.37–3.86 for compound 2, 78.7–91.3+/−1.61–2.92 for compound 3, and 68.6–74.5+/−2.04–4.81 for compound 4. A two-dimensional (2D) method for the separation and spectral characterization of the compounds consisted of development with mobile phase A, followed by B at 90° [91]. These methods were used to evaluate the quality of turmeric powders in India based on cucurmin content and the presence of extraneous colors. Branded samples ranged from 2.2% to 3.7% and loose powders from 0.3% to 2.6%. None of the branded powders contained artificial colors, while 17% of the loose powders showed extraneous color from metanil yellow in the range of 1.0–8.5 mg/g, which may pose health threats [92].

Curcumin, demethoxycurcumin, and bisdemethoxycurcumin from seven different germplasm of turmeric (*Curcuma longa* L.) were quantified by Soxhlet extraction, cleanup by gradient elution silica gel column chromatography, separation of samples and standards applied with a Linomat 5 on silica gel 60 G F aluminum plates in a TTC with chloroform-methanol (48:2) mobile phase, and absorption-reflection densitometry using a TLC Scanner 3 at 425 nm [12].

HPTLC-EDA-MS was demonstrated for the discovery and quantification of bioactive compounds, including colorant curcumin, in 21 ethanolic plant extracts.

Three selected bioassays were performed for EDA: *Aliivibrio fischeri* with the TLC Bioluminizer, *Bacillus subtillis* for microbial detection, and DPPH* for antioxidants. Bioactive components were confirmed in the extracts by use of the HPTLC-MS Interface with an ESI single quadrupole Advion expression mass spectrometer. For HPTLC, an ATS 4 applicator, silica gel 60 F plates, toluene-ethyl acetate-formic acid (9:6:0.4) mobile phase in a TTC, TLC Visualizer, TLC Scanner 3, and Chromatogram Immersion Device for derivatization with 1% ethanolic aluminum chloride reagent and application of EDA reagents were utilized [93].

### 4.5.5 Flavonoids

High-throughput screening of flavonoids in taro by HPTLC was done on 350 different accessions. Ten flavones were successfully detected in the corm; these are responsible for the attractive yellow color of the flesh and fibers. Silica gel 60 F plates, an ATS 4 applicator, an ADC 2 chamber for development with ethyl acetate-methanol-acetic acid-formic acid-water (30:1:2:1:3), and a TLC Scanner 4 with tungsten and deuterium lamps were employed to measure $R_f$ values and peak areas, in order to identify flavonoid bands and score tracks, based on the presence or absence of bands for comparison of varieties [94].

### 4.5.6 Vitamins

Cobalamin, commonly known as red-colored vitamin B12, was determined by miniaturized HPTLC on 5.0 × 7.5 cm Lichrospher silica gel 60 F sheets in animal-based foods such as meat, milk, and fish after purification of extracts by using a EASI-EXTRACT B12 immunoaffinity column [17]. Separation and identification of vitamin B12 were obtained with a 5-cm development in the dark by using 2-propanol-ammonia-water (7:1:2) or *n*-butanol-2-propanol-water (10:7:10) in less than 45 min, with an LOD of 34 ng at 254 nm.

## 4.6 CONCLUSIONS AND FUTURE PROSPECTS

Chemical structures of numerous synthetic dyes and carotenoids are pictured in some of the papers cited previously in this chapter [29,30,33,34,38,44,48,50,79,82].

TLC will continue to be widely used for qualitative and quantitative analyses related to the presence of synthetic and natural colorants in a wide variety of food and food-related matrices, as described in the literature cited previously in this chapter, plus, for example, identification and quantification of carotenoids in fruits [95] and yellow fleshed yams [96], rhodamine B and tartrazine in crackers (silica gel 60 F plates with isopropanol-ammonia [4:0.5] mobile phase) [97], prohibited azo Sudan dye para red in chili powder and tomato sauce samples extracted by homogenization with toluene [98], and adulterant copper chlorophyll (major component: Cu-pyropheophytin A) from slurry green dyes obtained from an edible-oil-processing factory [99]. These new applications will be driven in part by the recent confirmation that TLC is very powerful in the study of complex mixtures of synthetic dyes when linked to in situ reflectance spectrometry and SERS [100].

Use of natural colorants characterized by TLC in food products will become more acceptable owing to health considerations such as their nonallergic and non-carcinogenic properties [101]. TLC-EDA will be used more widely to evaluate natural food colorants in terms of their radical scavenging properties and general bioactivity, which give them the potential for medicinal value as well as a food additive for aesthetic purposes. EDA will be carried out by hyphenation of TLC with DPPH* and *Aliivibrio fischeri* tests, as well at the so-called "dot blot" DPPH* test, in which antioxidant potential is evaluated on spots applied to TLC plates without mobile phase development (nonchromatographic method). Although this concept to evaluate antioxidant activity of food colorants is carried out widely on undeveloped TLC plates, for example, the DPPH* "center dot" scavenging method, in which spot images are captured with an HP Deskjet scanner and evaluated with ImageJ software to determine sample concentrations providing 50% color reduction (CSC50) values, compared with standards [102], there are significant advantages when separations on the plate are coupled with bioassays, because correct identification of active compounds is more probable, and the mobile-phase organic solvents are completely evaporated before the applied biotest and cannot adversely affect it, as is possible in the case of HPLC-bioassay [103]. A recent book chapter reviewed the techniques and applications of the TLC-DPPH* antioxidant activity test, including natural food colorants [104]. It is also expected that additional studies will be dedicated to understanding, optimizing, and predicting dye separations, as was done on silica gel and cellulose layers with methanol-water and dichloromethane-water mobile phases in varying proportions, using quantum mechanics calculations and chemometrics [105].

It is unquestionable that food colorant analysis will be performed more often in future years by TLC coupled online with different modes of MS, using commercial and homemade interfaces for automated introduction of analytes from the plate into the mass spectrometer. This would be benefited by the use of Merck Millipore special purity HPTLC plates (20 × 10 cm, 100-μm layer thickness) designed for coupling with MS and TLC plates (20 × 20 cm, 200 μm) for coupling with MS and NMR spectrometry, which requires larger amounts of a substance. If the Office Chromatography concept becomes developed into a practical, commercial UTLC system, it will certainly be applied for "green" food colorant analyses.

Electrophoresis separations of selected synthetic colorants on glass-backed cellulose, polyamide, silica gel 60 W, RP18 W, and aluminum oxide layers were seen to be promising and may continue to be researched [106]. However, although these are planar separations, they are not chromatography, because analyte migration is caused by electrical forces rather than by flow of a mobile phase.

## ACKNOWLEDGMENTS

I thank Karen F. Haduck of the Lafayette College Interlibrary Loan Department for her invaluable work in obtaining copies of many of the publications cited in this chapter, without which it could not have been written.

## REFERENCES

1. Oka, H., Ozeki, N., Hayashi, T., and Itakura, Y., Analysis of natural colorings in foods by thin layer chromatography, *J. Liq. Chromatogr. Relat. Technol.,* 30, 2021–2036, 2007.
2. Kucharska, M. and Grabka, J., A review of chromatographic methods for determination of synthetic food dyes, *Talanta*, 80, 1045–1051, 2010.
3. Scotter, M.J., Methods for the determination of European Union permitted added natural colours in foods: A review, *Food Addit. Contam.,* 28, 527–596, 2011.
4. Hosu, A. and Cimpoiu, C., Thin layer chromatography applied in quality assessment of beverages derived from fruits, *J. Liq. Chromatogr. Relat. Technol.,* 40, 239–246, 2017.
5. Rovina, K., Siddiquee, S., and Shaarani, S.M., Toxicology, extraction and analytical methods for determination of amaranth in food and beverage products, *Trends Food Sci. Technol.*, 65, 68–79, 2017.
6. Stiefel, C. and Morlock, G.E., Application of hyphenated HPTLC in food, commodity and cosmetics analysis, HPTLC 2017, *International Symposium for High Performance Thin Layer Chromatography*, Berlin, Germany, 2017, Abstract O-10.
7. Rovina, F., Acung, L.A., Siddiquee, S., Akanda, J.H., and Shaarani, S. Md., Extraction and analytical methods for determination of sunset yellow (E110)—A review, *Food Anal. Methods,* 10, 773–787, 2017.
8. Ristivojevic, P., Milojkovic-Opsenica, D., and Morlock, G.E., Hyphenation of planar chromatography with chemometrics, HPTLC 2017, *International Symposium for High Performance Thin Layer Chromatography*, Berlin, Germany, July 4–8, 2017, Abstract O-54.
9. Yamjala, K., Nainar, M.S., and Ramisctti, N.R., Methods for analysis of azo dyes employed in food industry—A review, *Food Chem.,* 192, 813–824, 2016.
10. Pico, Y., Ultrasound assisted extraction for food and environmental samples, *TRAC Trends Anal. Chem.*, 43, 84–99, 2013.
11. Rovina, K., Siddiquee, S., and Shaarani, S.M., Extraction, analytical and advanced methods for detection of allura red AC (E129) in food and beverage products, *Front. Microbiol.,* 7, 798, 2016. doi:10.3389/fmicb.2016.00798.
12. Paramasivam, M., Poi, R., Banerjee, H., and Bandyopadhyay, A., High performance thin layer chromatographic method for quantitative determination of curcuminoids in *Curcuma longa* germplasm, *Food Chem.* 113, 640–644, 2009.
13. Santana, A.L., Johner, J.C.F., and Meireles, M.A.A., TLC profile of annatto extracts obtained with supercritical carbon dioxide and subsequently high pressure phase equilibrium data, *Food and Public Health*, 6, 15–25, 2015.
14. Santos, D.T., Albuquerque, C.L.C., and Meireles, A.A., Antioxidant dye and pigment extraction using a homemade pressurized solvent extraction system, *Procedia Food Sci*, 1, 1581–1588, 2011.
15. Rovina, K., Siddiquee, S., and Shaarani, S., A review of extraction and analytical methods for the determination of tartrazine (E 102) in foodstuffs, *Crit. Rev. Anal. Chem.*, 47, 309–324, 2017.
16. Asfaram, A., Ghaedi, M., and Goudarzi, A., Optimization of ultrasound assisted dispersive solid phase microextraction based on nanoparticles followed by spectrophotometry for the simultaneous determination of dyes using experimental design, *Ultrason. Sonochem.,* 32, 407–417, 2016.
17. Watanabe, F. and Bito, T., Determination of cobalamin and related compounds in foods, *J. AOAC Int.*, 101, 1308–2018.
18. Rathore, M., Khanam, A.J., and Gupta, V., Metal cation exchanged cellulose as new layer material for identification and separation of organic dyes, *Adv. Appl. Sci. Res.*, 6, 97–104, 2015.

19. Tuzimski, T. and Wozniak, A., Application of solid phase extraction and planar chromatography with diode array detection to the qualitative and quantitative analysis of dyes in beverages, *J. Planar Chromatogr.-Mod. TLC*, 21, 89–96, 2008.
20. Rabel, F. and Sherma, J., Review of the state of the art of preparative thin layer chromatography, *J. Liq. Chromatogr. Relat. Technol.*, 40, 165–176, 2017.
21. Clark, J.E. and Olesik, S.V., Electrospun glassy carbon ultra-thin layer chromatography devices, *J. Chromatogr. A*, 1217, 4655–4662, 2010.
22. Frolova, A.M., Konovalova, O.Y., Loginova, L.P., Bulgakova, A.V., and Boichenko, A.P., Thin layer chromatographic plates with a monolithic layer of silica: Production, physical-chemical characteristics, separation capabilities, *J. Sep. Sci.* 34, 16–17, 2011.
23. Jim, S.R., Taschuk, M.T., Morlock, G.E., Bezuidenhout, L.W., Schwack, W., and Brett, M.J., Engineered anisotropic microstructures for ultrathin layer chromatography, *Anal. Chem.*, 82, 5349–5356, 2010.
24. Rabel, F. and Sherma, J., A review of advances in two dimensional thin layer chromatography, *J. Liq. Chromatogr. Relat. Technol.*, 39, 627–639, 2016.
25. Stiefel, C. and Morlock, G.E., Separation of pigment formulations by HPTLC/AMD, HPTLC 2017, *International Symposium for High Performance Thin Layer Chromatography*, Berlin, Germany, 2017, Abstract P–10.
26. Fagundes, T.D.F., Dutra, K.D.B., Ribeiro, C.M.R., Epifanio, R.D., and Valverde, A.L., Using a sequence of experiments with turmeric pigments from foods to teach extraction, distillation, and thin layer chromatography to introductory organic chemistry students, *J. Chem. Educ.*, 93, 326–329, 2016.
27. McCullagh, J.V. and Ramos, N., Separation of the carotenoid bixin from annatto seeds using thin layer and column chromatography, *J. Chem. Educ.*, 85, 948–950, 2008.
28. Tami, K., Popova, A., and Proni, G., Engaging students in real world chemistry through synthesis and confirmation of azo dyes via thin layer chromatography to determine the dyes present in everyday foods and beverages, *J. Chem. Educ.*, 94, 471–475, 2017.
29. De Andrade, F.I., Guedes, M.I.F., Vieira, I.G.P., Mendes, F.N.P., Rodrigues, P.A.S., Maia, C.S.C., Avila, M.M.M., and Ribeiro, L.D., Determination of synthetic food dyes in commercial soft drinks by TLC and ion pair HPLC, *Food Chem.*, 157, 193–198, 2014.
30. Gierak, A., Skorupa, A., and Lazarska, I., Capillary action liquid chromatography: New chromatographic technique for separation and determination of colour substances, *Adsorpt. Sci. Technol.*, 33, 639–643, 2015.
31. Izzo, F.C., Vitale, V., Fabbro, C., and Van Keulen, H., Multi-analytical investigation on felt tip pen inks: Formulation and preliminary photo-degradation study, *Microchem. J.*, 124, 919–928, 2016.
32. Komissarchik, S. and Nyanikova, G., Test systems and a method for express detection of synthetic food dyes in drinks, *LWT – Food Sci. Technol.*, 58, 315–320, 2014.
33. Wlodarczyk, E. and Zarzycki, P.K., Chromatographic behavior of selected dyes on silica and cellulose micro-TLC plates: Potential application as target substances for extraction, chromatographic, and/or microfluidic systems, *J. Liq. Chromatogr. Relat. Technol*, 40, 259–281, 2017.
34. Tatebe, C., Zhong, X., Ohtsuki, T., Kubota, H., Sato, K., and Akiyama, H., A simple and rapid chromatographic method to determine unauthorized basic colorants (rhodamine B, auramine O, and pararosaniline) in processed foods, *Food Sci. Nutr.*, 2, 547–556, 2014.
35. Zahra, N., Alim-un-Nisa, Fatima, Z., Kalim, I., and Saeed, K. Identification of synthetic food dyes in beverages by thin layer chromatography, *Pak. J. Food Sci.*, 25, 178–181, 2015.
36. Mohammad, A., Khan, M., Mobin, R., and Mohammad, F., A new thin layer chromatographic system for the identification and selective separation of brilliant blue food dye – application of a green solvent, *J. Planar Chromatogr.-Mod. TLC*, 29, 446–452, 2016.

37. Halka-Grysinska, A., Gwarda, R.L., Pawelek, K., Baj, T., and Dzido, T.H., Reversed phase stepwise gradient thin layer chromatography of 19 test dye mixtures with a single void volume of the mobile phase, *J. Planar Chromatogr.-Mod. TLC*, 30. 113–120, 2017.
38. Tang, T.X., Xu, X.J., Wang, D.M., Zhao, Z.M., Zhu, L.P., and Yang, D.P., A rapid and green limit test method for five synthetic colorants in foods using polyamide thin layer chromatography, *Food Anal. Methods*, 8, 459–466, 2015.
39. Zhang, Y., Surface enhanced Raman spectroscopy (SERS) combined techniques for high performance detection and characterization, *TRAC Trends Anal. Chem.*, 90, 1–13, 2017.
40. Gao, F., Hu, Y.X., Chen, D., Li-Chen, E.C.Y., Grant, E., and Lu, X.N., Determination of sudan I in paprika powder by molecularly imprinted polymers-thin layer chromatography-surface enhanced Raman spectroscopic biosensor, *Talanta*, 143, 344–352, 2015.
41. Wang, C., Cheng, F., Wang, Y., Gong, Z., Fan, M., and Hu, J., Single point calibration for semi-quantitative screening based on an internal reference in thin layer chromatography-SERS: The case of rhodamine B in chili oil, *Anal. Methods*, 6, 7218–7223, 2014.
42. Kong, X., Squire, K., Chong, X., and Wang, A.X., Ultra-sensitive lab-on-a-chip detection of sudan I in food using plasmonics-enhanced diatomaceous thin film, *Food Control*, 79, 258–265, 2017.
43. Zhang, B.B., Shi, Y., Chen, H., Zhu, Q.x., Lu, F., and Li, Y.w., A separable surface enhanced Raman scattering substrate modified with MIL–101 for detection of overlapping and invisible compounds after thin layer chromatography development, *Anal. Chim. Acta*, 997, 35–43, 2018.
44. Cheng, S.-C., Huang, M.-Z., and Shiea, J., Thin layer chromatography/laser induced acoustic desorption/electrospray ionization mass spectrometry, *Anal. Chem.*, 81, 9274–9281, 2009.
45. Lin, S..Y., Huang, M.-Z., Chang, H.-C., and Shiea, J., Using electrospray assisted laser desorption/ionization mass spectrometry to characterize organic compounds separated on thin layer chromatography plates, *Anal. Chem.*, 79, 8789–8795, 2007.
46. Morlock, G.E. and Brett, N., Correct assignment of lipophilic dye mixtures? A case study for high performance thin layer chromatography-mass spectrometry and performance data for the TLC-MS interface, *J. Chromatogr. A*, 1390, 103–111, 2015.
47. Zhang, J., Zhou, Z., Yang, J., Zhang, W., Bai, Y., and Liu, H., Thin layer chromatography/plasma assisted multiwavelength laser desorption ionization mass spectrometry for facile separation and selective identification of low molecular weight compounds, *Anal. Chem.*, 84, 1496–1503, 2012.
48. Soponar, F., Mot, A.C., and Sarbu, C., Quantitative determination of some food dyes using digital processing of images obtained by thin layer chromatography, *J. Chromatogr. A*, 1188, 295–300, 2008.
49. Kandler, H., Bleisch, M., Widmer, V., and Reich, E., A validated HPTLC method for the determination of illegal dyes in spices and spice mixtures, *J. Liq. Chromatogr. Relat. Technol.*, 32, 1273–1288, 2009.
50. Morlock, G.E. and Oellig, C., Rapid planar chromatographic analysis of 25 water soluble dyes used as food additives, *J. AOAC Int.* 92, 745–756, 2009.
51. Tuzimski, T., Determination of sulfonated water soluble azo dyes in food by SPE coupled with HPTLC-DAD, *J. Planar Chromatogr.-Mod. TLC*, 24, 281–289, 2011.
52. Cassoni, D., Boldan, A., and Cobzac, S.C., TLC-densitometric determination of synthetic food colorants from pharmaceutical powders, *Studia UBB Chemia,* 57, 83–92, 2012.
53. Cobzac, S.C., Casoni, D., and Pop, D., Tartrazine determination from mustard sample by TLC-photodensitometry and TLC-digital processing of images, *J. Planar Chromatogr. Mod. TLC*, 25, 542–547, 2012.

54. Cobzac, S.C., Casoni, D., Fazakas, A.L., and Sarbu, C., Determination of food synthetic dyes in powders for jelly desserts using slit scanning densitometry and image analysis methods, *J. Liq. Chromatogr. Relat. Technol.*, 35, 1429–1443, 2012.
55. Vlajkovic, J., Andric, P., Ristivojevic, P., Radoici, A., Tesic, Z., and Opsenica, D., Development and validation of a TLC method for the analysis of synthetic foodstuff dyes, *J. Liq. Chromatogr. Relat. Technol.,* 36, 2476–2488, 2013.
56. Habe, T.T. and Morlock, G.E., Office chromatography: Precise printing of sample solutions on miniaturized thin layer phases and utilization for scanning direct analysis in real time mass spectrometry, *J. Chromatogr. A,* 1413, 127–134, 2015.
57. Kanyai, S.S., Habe, T.T., Cushman, C.V., Dhunna, M., Roychowdhury, T., Farnsworth, P.B., Morlock, G.E., and Linford, M.R., Microfabrication, separations, and detection by mass spectrometry on ultrathin-layer chromatography plates prepared via the low pressure chemical vapor deposition of silicon nitride onto carbon nanotube templates, *J. Chromatogr. A*, 1404, 115–123, 2015.
58. Wannenmacher, J., Jim, S.R., Taschuk, M.T., Brett, M.J., and Morlock, G.E., Ultrathin-layer chromatography on $SiO_2$, $Al_2O_3$, $TiO_2$, and $ZrO_2$ nanostructured thin films, *J. Chromatogr. A,* 1318, 234–243, 2013.
59. Niamiang, P., Supaphol, P., and Morlock, G.E., Performance of electrospun polyacrylonitrile nanofibrous phases, shown for the separation of water soluble food dyes via UTLC-Vis-ESI-MS, *Nanomaterials*, 7, 8, 2017. doi:10.3390/nano7080218.
60. Sobanska, A.W., Pyzowski, J., and Brzezinska, E., SPE/TLC/densitometric quantification of selected synthetic food dyes in liquid foodstuffs and pharmaceutical preparations, *J. Anal. Methods*, 2017, 1–9, 2017. doi:10.1155/2017/9528472.
61. Milz, B., Schnurr, P., Grafmueller, J., Oehler, K., and Spangenberg, B., A validated quantification of sudan red dyes in spicery using TLC and a 16 bit flatbed scanner, *J. AOAC Int.*, in press, 2018.
62. Djelal, H., Cornee, C., Tartivel, R., Lavastre, O., and Abdeltif, A., The use of HPTLC and direct analysis in real time-of-flight mass spectrometry (DART-TOF-MS) for rapid analysis of degradation by oxidation and sonication of an azo dye, *Arab. J. Chem.,* 10, S1619-S1628, 2017.
63. Viral, S. and Kunjal, P., Degradation of azo dye acid blue 158 by soil microbes, *Res. J. Biotechnol.* 7, 50–57, 2012.
64. Valandoostarani, S., Lotfabad, T.B., Heidannasaab, A., and Yaghmaei, S., Degradation of azo dye methyl red by *Saccharomyces cerevisiae* ATCC 9873, *Int. Biodeterior. Biodegrad.* 125, 62–72, 2017.
65. Pathak, H. and Madamwar, D., Biosynthesis of indigo dye by newly isolated naphthalene-degrading strain *Pseudomonas* sp HOB1 and its application in dyeing cotton fabric, *Appl. Biochem. Biotechnol.*, 160, 1616–1626, 2010.
66. Koli, S.H., Suryawanshi, R.K., Patil, C.D., and Patil, S.V., Fluconazole treatment enhances extracellular release of red pigments in the fungus *Monascus purpureus*, *FEMS Microbiol. Lett.*, 364, 2017, doi: 10.1093/femsle/fnx058.
67. Li, Y., Yang, Y., Yin, S., Zhou, C., Ren, D., and Sun, C., Inedible azo dyes and their analytical methods in foodstuffs and beverages, *J. AOAC Int.*, 101, 1314–1327, 2018, doi: 10.5740/jaoacint.18-0048.
68. Cimpoiu, C., Hosu, A., Briciu, R., and Miclaus, V., Monitoring the origin of wine by reversed phase thin layer chromatography, *J. Planar Chromatogr.-Mod. TLC,* 20, 407–410, 2007.
69. Kim, D.M., Bae, J.S., Lee, D.S., Lee, H., Joo, M.H., and Yoo, S.H., Positive effects of glycosolated anthocyanin isolated from an edible berry fruit (*Acanthopanax sessiliflorum*) on its antioxidant activity and color stability, *Food Res. Int.*, 44, 2258–2263, 2011.

70. Filip, M., Vlassa, M., Copaciu, F., and Coman, V., Identification of anthocyanins and anthocyanidins from berry fruits by chromatographic and spectroscopic techniques to establish the juice authenticity from market, *J. Planar Chromatogr.Mod. TLC*, 25, 534–541, 2012.
71. Boehmdorfer, S., Oberlerchner, J., Rosenau, T., and Grausgruber, H., Clustering analysis of colored wheat varieties by anthocyanin patterns, HPTLC 2017, *International Symposium for High Performance Thin Layer Chromatography*, Berlin, Germany, July 4–8, 2017, Abstract O–56.
72. Krueger, S. and Morlock, G.E., Anthocyanin profiles of colored wheat crosses via HPTLC, HPTLC 2017, *International Symposium for High Performance Thin Layer Chromatography*, Berlin, Germany, July 4–8, 2017, Abstract P–31.
73. Lago, C., Landoni, M., Cassani, E., Cantaluppi, E., Doria, E., Nielsen, E., Giorgi, A., and Pilu, R., Study and characterization of ancient European flint white maize rich in anthocyanins: Millo Corvo from Galacia, *PLoS One,* 10, 5, 2015. doi:10.1371/journal.pone.0126521, 2015.
74. Lago, C., Cassani, E., Zanzi, C., Landoni, M., Trovato, R., and Pilu, R., Development and study of a maize cultivar rich in anthocyanins: Coloured polenta, a new functional food, *Plant Breeding,* 133, 210–217, 2014.
75. Lebot, V., Michalet, S., and Legendre, L., Identification and quantification of phenolic compounds responsible for the antioxidant activity of sweet potatoes with different flesh colours using high performance thin layer chromatography (HPTLC), *J. Food Compos. Anal.,* 49, 94–101, 2016.
76. Skorek, M., Pytlakowska, K., Sajewicz, M., and Kowalska, T., Thin layer chromatographic investigation of plant pigments in selected juices and infusions of cosmetological importance and their antioxidant potential, *J. Liq. Chromatogr. Relat. Technol.* 40, 311–319, 2017.
77. Krueger, S., Urmann, O., and Morlock, G.E., Development of a planar chromatographic method for quantitation of anthocyanes in pomace, feed, juice, and wine, *J. Chromatogr. A,* 1289, 105–118, 2013.
78. Lata, E., Fulczyk, A., Kowalska, T., and Sajewicz, M., Thin layer chromatographic method of screening the anthocyanes containing alimentary products and precautions taken at the method development step, *J. Chromatogr. A*, 1530, 211–218, 2017.
79. Briciu, R.D., Casoni, D., and Bischin, C., Thin layer chromatography separation of some carotenoids, retinoids, and tocopherols, *Studia UBB Chemia,* 53, 67–75, 2008.
80. Agatonovic-Kustrin, S., Morton, D.W., and Ristivojevic, P., Assessment of antioxidant activity in Victorian marine algal extracts using high performance thin layer chromatography and multivariate analysis, *J. Chromatogr. A,* 1468, 228–235, 2016.
81. Fu, H.F., Xie, B.J., Ma, S.J., Zhu, X.R., Fan, G., and Pan, S.Y., Evaluation of antioxidant activities of principal carotenoids available in water spinach (Ipomoea aquatica), *J. Food Compos. Anal.*, 24, 288–297, 2011.
82. Rodic, Z., Simonovska, B., Albreht, A., and Vovk, I., Determination of lutein *by* high performance thin layer chromatography using densitometry and screening of major dietary carotenoids in food supplements, *J. Chromatogr. A*, 1231, 59–65, 2012.
83. Starek, M., Guja, A., Dabrowska, M., and Krzek, J., Assay of β-carotene in dietary supplements and fruit juices by TLC-densitometry, *Food Anal. Methods*, 8, 1347–1355, 2015.
84. Milz, B., Minar, Y.A., and Spangenberg, B., Quantification of astaxanthin in salmons by chemiluminescence and absorption after TLC separation, *J. Liq. Chromatogr. Relat. Technol.*, 4, 358–363, 2018.
85. Zeb, A. and Murkovic, M., High performance thin layer chromatographic method for monitoring the thermal degradation of β-carotene in sunflower oil, *J. Planar Chromatogr. Mod. TLC,* 23, 35–39, 2010.

86. Zeb A. and Murkovic, M., Pro-oxidant effects of b-carotene during thermal oxidation of edible oils, *J. Am. Oil Chem. Soc.,* 90, 881–889, 2013.
87. Kiani, S., Minaei, S., and Ghasemi-Varnamkhasti, M., Instrumental approaches and innovative systems for saffron quality control, *J. Food Eng.*, 216, 1–10, 2018.
88. Djozan, D., Karimian, G., Jouyban, A., Iranmanesh, F., Gorbanpour, H., and Alizadeh-Nabil, A.A., Discrimination of saffron based on thin layer chromatography and image analysis, *J. Planar Chromatogr.Mod. TLC*, 27, 274–280, 2014.
89. Seresthi, H., Poursorkh, Z., Aliakbarzadeh, G., Zarre, S., and Ataolahi, S., An image analysis of TLC patterns for quality control of saffron based on soil salinity effect: A strategy for data (pre)-processing, *Food Chem.*, 239, 831–839, 2018.
90. Havlikova, L., Satinsky, D., Opletal, L., and Solich, P., A fast determination of chlorophylls in barley grass juice powder using HPLC fused core column technology and HPTLC, *Food Anal. Methods,* 7, 629–635, 2014.
91. Dixit, S., Khanna, S.K., and Das, M., A simple 2-directional high performance thin layer chromatographic method for simultaneous determination of curcumin, metanil yellow, and sudan dyes in turmeric, chili, and curry powders, *J. AOAC Int.*, 91, 1387–1396, 2008.
92. Dixit, S., Purshottam, S.K., Khanna, S.K., and Das, M., Surveillance of the quality of turmeric powders from city markets of India on the basis of curcumin content and the presence of extraneous colours, *Food Addit. Contam.*, 26, 1227–1231, 2009.
93. Taha, M.N., Krawinkel, M.B., and Morlock, G.E., High performance thin layer chromatography linked with (bio)assays and mass spectrometry—A suited method for discovery and quantification of bioactive components? Exemplarily shown for turmeric and milk thistle, *J. Chromatogr. A,* 1394, 137–147, 2015.
94. Lebot, V., Lawac, F., Michalet, S., and Legendre, L., Characterization of taro [*Colocasia esculenta* (L.) Schott] germplasm for improved flavonoid composition and content, *Plant Genet. Resour. C.*, 15, 260–268, 2017.
95. de Lanerolle, M.S., Priyadarshani, A.M.B., Sumithraarachchi, D.B., and Jansz, E.R., The carotenoids of *Pouteria campechiana* (sinhala: *Ratalawulu*), *J. Natl. Sci. Found. Sri Lanka*. 36, 95–98, 2008.
96. Ferede, R., Maziya-Dixon, B., Alamu, O.E., and Asiedu, R., Identification and quantification of major carotenoids of deep yellow fleshed yam (tropical *Dioscorea dumetorum*), *J. Food Agric. Environ.*, 8, 160–166, 2010.
97. Levita, J., Megantara, S., and Mutakin, Photometric titration method to determination bromination of red and yellow dyes in crackers, *Int. J. Chem., (Toronto, ON, Canada)*, 4, 80–85, 2012.
98. Mustafa, S., Khan, R.A., Sultana, I., Nasir, N., and Tariq, M., Estimation of para red dye in chili powder and tomato sauces by a simple spectrophotometric method followed by thin layer chromatography, *J. Appl. Sci. Environ. Manage.* 17, 177–184, 2013.
99. Peng, G., Chang, M., Fang, M., Liao, C., Tsai, C., Tseng, S., Kao, Y., Chou, H., and Cheng, H., Incidents of major food adulteration in Taiwan between 2011 and 2015, *Food Control*, 72, 145–152, 2017.
100. Germinario, G., Garrappa, S., D'Ambrosio, V., van der Werf, D., and Sabbatini, L., Chemical composition of felt tip inks, *Anal. Bioanal. Chem.,* 410, 1079–1094, 2018.
101. Jadhav, B.A. and Joshi, A.A., Extraction and quantitative estimation of bio-active component (yellow and red carthamin) from dried safflower petals, *Indian J. Sci. Technol.*, 8 (16), 1–5, 2015.
102. Akar, Z., Kucuk, M., and Dogan, H., A new colorimetric DPPH center dot scavenging activity method with no need for a spectrophotometer applied on synthetic and natural antioxidants and medicinal herbs, *J. Enzyme Inhib. Med. Chem.*, 32, 640–647, 2017.

103. Weiss, S.C., Egelenmeyer, N., and Schulz, W., Coupling of in vitro bioassays with planar chromatography in effect-directed analysis, *Adv, Biochem. Eng. Biotechnol.*, 157, 187–224, 2017.
104. Sherma, J., Studies on the antioxidant activity of foods and food ingredients by thin layer chromatography-direct bioautography with 2,2'-diphenyl–1-picrylhydrazyl radical (DPPH*), *Adv. Chromatogr.*, 55, 229–244, 2018.
105. Pereira, J.C., Marques, J.M.C., Wlodarczyk, E., Fenert, B., and Zarzycki, P.K., Towards the understanding of micro-TLC behavior of various dyes on silica and cellulose stationary phases using data mining approach, *J. AOAC Int.*, in press, 2018.
106. Lewandowska, L., Wlodarczyk, E., Fenert, B., Kaleniecka, A., and Zarzycki, P.K., A preliminary study for the fast prototyping of simple electroplanar separation systems based on various natural polymers and planar chromatographic stationary phases, *J. Planar Chromatogr. Mod. TLC.*, 30, 440–452, 2017.

# 5 Capillary Electrophoresis Coupled to Mass Spectrometry as a Powerful Tool to Investigate Heterogeneities, Conformers, and Oligomers of Intact Proteins

*Yannis-Nicolas François, Anne-Lise Marie, Coralie Ruel, Rabah Gahoual, Nguyet Thuy Tran, and Myriam Taverna*

## CONTENTS

## 5.1 INTRODUCTION

The introduction of electrospray (ESI) and matrix-assisted laser desorption ionization (MALDI) in the 1980s has successfully addressed the transfer of ions from biological molecules such as proteins to the gas phase. These ionization processes have offered an entirely unexplored field to mass spectrometry (MS), represented by the analysis of biological compounds and/or samples. Thus, more than 30 years later, we are still experiencing the outcome of these major breakthroughs.

Consequently, the implementation of separation techniques prior to MS analysis has been investigated in order to enable, for example, the characterization of complex samples. In this context, capillary electrophoresis (CE) hyphenated with MS (CE-MS) has been rapidly considered as particularly relevant for several reasons. In CE separation, the mobility of a compound is partially determined by the charge in solution, which is mainly influenced by the background electrolyte (BGE's) pH and the hydrodynamic radius of the compound. The mobility differences between the analytes composing the sample lead to their separation. For intact proteins, the selectivity provided by CE appears interesting and complementary to that obtained with chromatographic methods; this drives its implementation in conjunction with MS. The separation of proteins has been achieved for decades using electrokinetic separation such as conventional gel electrophoresis. CE represents the most performant electrophoretic separation format in terms of resolution, separation efficiency, and peak capacity.

From an MS prospect, CE selectivity is completely orthogonal to *m/z* measurements provided by MS analysis, which strongly suggests a suitable complementarity between the two techniques. In addition, ESI and MALDI ionization yields are constant but limited. The introduction of a high quantity of ionizable compounds such as salts could reduce the sensitivity by ion suppression effect. Owing to the miniaturized format of CE instrumentation, BGE flow rates in the nanoliter range are involved for the separation, which could provide an optimal ionization efficiency of the sample content, resulting in a maximization of MS sensitivity.

In order to hyphenate CE directly to MS instrumentation, the principal requirement resides in maintaining the electrical field necessary to the electrophoretic separation and thus transferring the capillary outlet into the MS source. Over the past decades, different types of interface have been designed to enable CE coupling with ESI- and MALDI-MS instruments. In order to provide a robust coupling, the first generation of CE-MS interfacing had to compromise heavily on sensitivity, and this prevented the achievement of optimal CE-MS coupling performance. Nevertheless, CE-MS systems have recently benefited from major instrumental improvements, which allowed to enhance the compatibility between both techniques. These improvements proved to be particularly significant in the case of CE-ESI-MS coupling.

Concomitantly, the analysis of intact proteins has demonstrated a growing interest in various research fields such as the study of non-covalent complexes and the characterization of biopharmaceutical products. Intact proteins represent complex macromolecules with an extended range of potential heterogeneities. Microheterogeneity represents small variations compared with the global structure of the protein; such is the case of post-translational modifications (PTMs) such as glycosylation, acetylation, phosphorylation, sulfonation, and citrullination. These PTMs have been often identified to play a key role in *in vivo* signaling. The presence of conformational protein isomers is another source of heterogeneity, implying a difference only in the hydrodynamic radius but not in the charge. Heterogeneity can also arise from major alterations of the protein such as denaturation, misfolding dimerization, oligomerization, and aggregation. All these variations may modify the physicochemical properties or biological activity of the protein in a drastic manner. Therefore, it is important to develop analytical methodologies able to finely characterize these modifications, without complex dissociation induced in the ionization source, mainly in the case of dimers, oligomers, or aggregates. The specific selectivity provided by the electrophoretic separation, in addition to the characteristics of MS analysis, with possible access to structural information, has showed that CE-MS analysis is particularly relevant, with the opportunity to separate proteins exhibiting only faint differences.

This chapter aims at offering a comprehensive and critical insight on the key aspects of CE-MS analysis of intact proteins, from a method development and an application point of view. If CE-MS has already fully demonstrated its potential for digested proteins analysis, intact protein analysis remains an area to be explored and that constitutes the main focus of this chapter. First, the different types of instrumental designs developed in order to perform CE-MS hyphenation are discussed, with the objective to highlight the specificities and limitations of each type of CE-MS interface. The analysis of intact proteins using CE has demonstrated dramatic adsorption of proteins on the capillary walls when using bare fused silica capillaries, especially owing to electrostatic interactions. To alleviate that limitation, semi-permanent or covalent coatings are applied to the capillary walls in order to prevent protein adsorption during the separation. The major types of coatings compatible with CE-MS experiments are then described to emphasize their chemical nature and impact on the electrophoretic separation. Because of the reduced volume of CE capillary, most often, the quantity of injected sample is not sufficient for providing a good MS sensitivity. To maximize the loading of the sample, a preconcentration step may be implemented offline or online. The most important strategies used in CE-MS analysis for sample preconcentration are described herein. The applications reported show that CE-MS analysis could be successfully implemented to study intact proteins over the different facets defining their structure. That versatility is illustrated in this chapter, with the application of CE-MS to the study of protein aggregates (formation and full characterization) and protein conformational changes. From another angle, the other applications described demonstrate the possibility to perform a detailed analysis of the major types of PTMs in order to study natural proteoforms or potential chemical degradations in an extensive manner.

## 5.2 TECHNICAL AND PRACTICAL ASPECTS

### 5.2.1 Principles and Different Types of Interfaces

The coupling of high-resolution CE separations with high-sensitivity and selectivity MS detections represents a relevant approach for targeted or comprehensive analyses of intact proteins. Since the end of the 1980s, a large number of CE-MS interfaces have been developed, especially with two major MS sources: (i) ESI and (ii) MALDI. Basically, the development of CE-MS interfaces proceeds in two different ways, using sheath-liquid or sheathless devices. Proving the interest of CE-MS coupling for analytical applications, CE-MS interfaces are in constant development and are marketed under different forms. In these past decades, several reviews and a recent book were published to describe and discuss deeply the principles and fundamentals of CE-MS interface strategies [1–10]. The following subsection focuses on the description of the different CE-MS interfaces that have proved their usefulness for the characterization of intact proteins.

#### 5.2.1.1 CE-ESI-MS Interfaces

##### *5.2.1.1.1 Sheath-Liquid CE-ESI-MS Interfaces*

For the last three decades, the sheath-liquid interface has been the most commonly used CE-ESI-MS interface. The first development of this interface has been described by the group of Smith in 1987 [11]. On this type of interface, the output of CE capillary is positioned in a specially designed nebulizer (Figure 5.1a). The electrical contact is maintained by means of an additional liquid named sheath-liquid. This liquid circulates in the nebulizer coaxially with CE capillary thanks to pneumatic assistance to continuously inject the additional liquid (5–10 μL/min). A junction is formed at the end of the CE capillary between the sheath-liquid and the BGE. The electrical contact is maintained at this junction, because sheath-liquid is connected to the CE output electrode. As conventionally in ESI, a nebulization gas is used to facilitate the formation of a stable spray.

Therefore, sheath-liquid must be a conductive liquid. In some applications, the sheath-liquid has the capacity to promote ionization. A special attention is therefore paid to adjust the composition of the additional liquid to obtain the best performances [12]. The other advantage of the sheath-liquid is that it renders possible substantial improvement of the tolerance of MS to certain non-volatile BGEs, such as borate and phosphate buffers, commonly used for CE separation of proteins but with high ESI-MS interference [13].

The major drawback of this type of interface paradoxically comes from the sheath-liquid itself. Indeed, the flow of additional liquid is much greater than the BGE created by the electroosmotic flow (EOF). Thereby, some effects of "dilution" of the analytes are observed during ESI ionization. As a result, the sensitivity

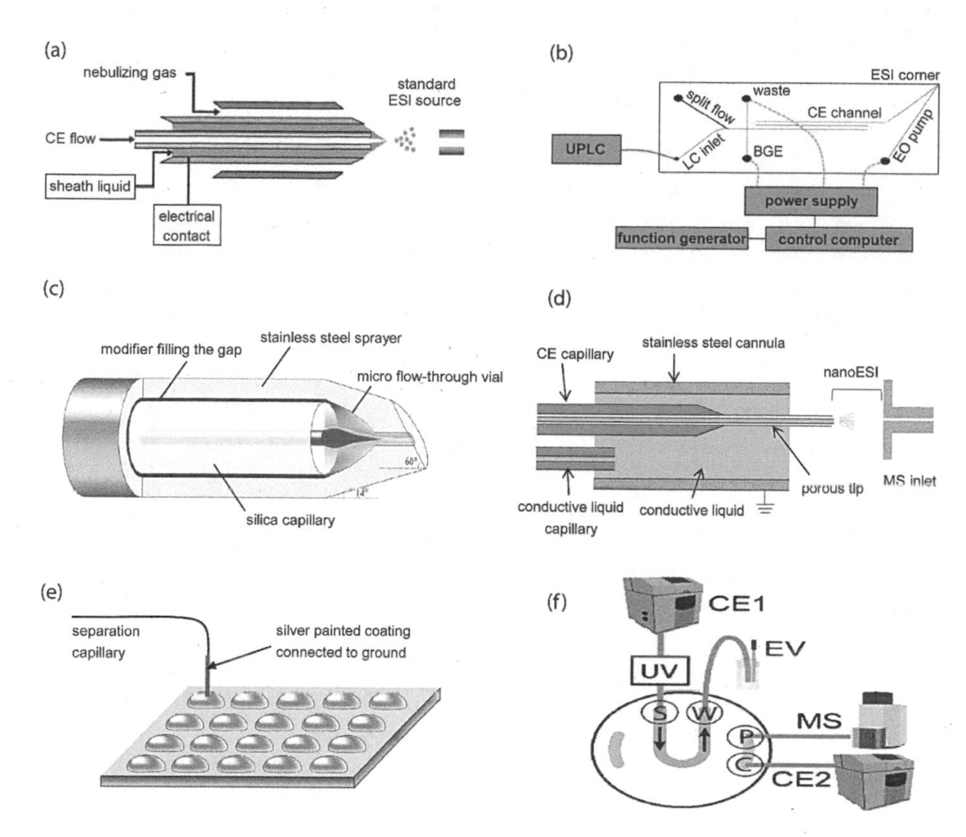


**FIGURE 5.1** Schematic of CE-MS interfaces: (a) sheath-liquid CE-ESI-MS, (b) chip-based CE-ESI-MS, (c) junction liquid CE-ESI-MS, (d) sheathless CE-ESI-MS, (e) sheath-liquid CE-MALDI-MS, and (f) 2D CE-CE-ESI-MS. ([a] Reprinted from *J. Chromatogr. A*, 1159, Haselberg, R. et al., Capillary electrophoresis–mass spectrometry for the analysis of intact proteins, 81–109, Copyright 2007, with permission from Elsevier; [b–e] Reprinted from Mellors, J. S. et al., *Anal. Chem.*, 85, 4100–4106, 2013, Zhong, X.F. et al., *Anal. Chem.*, 83, 4916–4923, 2011, Biacchi, M. et al., *Anal. Chem.*, 87, 6240–6250, 2015, Busnel, J.-M. et al., *Anal. Chem.*, 81, 3867–3872, 2009. Copyright (2013), (2011), (2015) and (2009) American Chemical Society. With permission; [f] Reprinted from Joos, K. et al., Two-dimensional capillary zone electrophoresis–mass spectrometry for the characterization of intact monoclonal antibody charge variants, including deamidation products, *Anal. Bioanal. Chem.*, 409, 2017, 6057–6067. With permission from Springer Nature.)

obtained with this type of interface is relatively limited. Haselberg et al. reported limits of detection (LODs) between 33 nM and 106 nM for four model proteins using sheath liquid CE-ESI-MS interface [14]. There is thus some ambivalence between the miniaturized character of CE, which induces a limited quantity of injectable sample, and the use of a high additional liquid flow rate. The historical sheath-flow interface, commercialized by Agilent Technologies (Waldbronn, Germany), was used in a large number of applications, including the separation of intact proteins [15–17].

Since the beginning of the twenty-first century, significant progresses have been made in the area of sheath-liquid interfacing. For example, in 2010, Dovichi's group developed an electrokinetically pumped sheath-flow interface. In their design, the separation capillary was placed inside a tapered glass emitter. Typical sheath-liquid flow rate, driven by electroosmosis, which is produced by the zeta potential, could be between 100 nL/min and 1000 nL/min [18]. This interface has several advantages compared with the traditional sheath-flow interfaces, including reduced sample dilution owing to a very low flow rate of sheath-liquid generated by electrokinetic forces, elimination of mechanical pumps, and stable operation in the nanoelectrospray regime. The performance of the electrokinetically pumped sheath-flow interface has been evaluated for a large number of applications, especially in the field of intact protein separation, with the development of a methodology to perform fast top-down intact protein characterization [19]. Recently, CMP Scientific (Brooklyn, USA) has commercialized this interface as the EMASS-II ion source. Evaluation of this interface has been performed, and it provides promising performances for intact protein analysis [20].

Microfabricated devices can also be considered as CE-ESI-MS interfaces, even if the geometry of the CE system is totally different. Indeed, using chip-based CE-ESI-MS, additional liquid is used to obtain stable sprays. ESI ionization is performed at one end of the microchip according to different strategies, such as the use of an ESI needle [21], or directly from the edge of a chip, with the spray potential applied between a reservoir and the inlet of the mass spectrometer [22]. When the classical CE-ESI-MS interfacing requires capillary length above 60 cm owing to CE instrument properties, the miniaturization of chip-based CE-ESI-MS allows to perform rapid separation with ultra-low sample volumes and can be customized in terms of channel lengths for specific applications [7]. Over the past decades, various approaches in the field of chip-based CE-ESI-MS devices have been described in the literature [8,9,23–26]. In 2008, the team of Ramsey published a fully integrated glass microfluidic device for performing high-efficiency CE-ESI-MS [27]. A thin-glass microchip corner served as an ESI emitter (Figure 5.1b). CE separation channel ends at the 90° corner of the chip and generates a flow out of the device by the incorporation of an electroosmotic pump. In 2013, the same group updated the system as an online hybrid multidimensional microfluidic system for comprehensive online liquid chromatography-CE-ESI-MS [28]. Major papers in the field of biotherapeutics characterization at the intact level were published these last years, proving the performance of the system for the separation and characterization of intact proteins [29,30]. Recently, a commercially version of this interface, named ZipChip™, is proposed by 908 Devices Inc. (Boston, USA).

##### *5.2.1.1.2 Liquid Junction CE-ESI-MS Interfaces*

Liquid junction interfaces retain some of the characteristics of sheath-liquid interfaces. The electrical contact is also maintained by means of an additional liquid. The development of liquid junction interfaces aims to preserve the advantages provided by additional liquids (favored ionization and stable ESI), while minimizing its dilution effect. This kind of interface was introduced by Lee et al. in 1988 [35]. Several designs of liquid junction interfaces have since been developed [4], their main difference being how to make the junction between the BGE and the additional liquid.

At its origin, the liquid junction interface was designed based on the use of a T-junction for the additional liquid to be brought in contact with the end of the CE capillary. This T-junction provides the connection between the CE capillary, the additional liquid, and a needle used to nebulize the liquid in the ESI source [36]. As in the case of sheath-liquid interfaces, the additional liquid is connected to the CE output electrode to maintain the electrical contact. Inside the junction, the distance between the CE capillary and the ESI needle is minimized to avoid peak broadening effects (typically between 25 μm and 50 μm). By maintaining the electrical contact in this manner, the additional liquid flow rate is significantly reduced to a few hundred nL/min. The direct consequence is that it is not necessary to use a nebulizing gas with this CE-ESI-MS interface [37]. The use of this interface enables to significantly reduce the effects of dilution comparatively to those observed with the sheath-liquid interface. Moreover, the separation between the CE capillary and the ESI needle facilitates its use because of the possibility to change them independently. However, the use of a T-junction creates dead volumes in the system, which tend to disrupt the flat front of the EOF and decrease the separation efficiency due to peak widening.

Recent improvements of liquid junction interfaces have been described by the group of Chen, with the development of a more physically robust junction-at-the-tip interface [31]. In this design, CE capillary is inserted entirely into a tapered stainless-steel hollow ESI emitter. The needle is filled with an additional liquid, connected to the CE output electrode. The small volume between the capillary end and the inner wall of the needle electrode tip forms a flow-through micro-vial that completes the CE setup requirement by acting as both the outlet vial and the terminal electrode (Figure 5.1c). During the separation, the additional liquid is injected at a low flow rate, around 300 nL/min, through the CE hydrodynamic injection. To promote the formation of small-diameter droplets, the design of the ESI needle is modified, in order to be beveled. Chen's group reported the evaluation of this interface for intact protein analysis [31].

#### 5.2.1.1.3 *Sheathless CE-ESI-MS Interfaces*

Sheathless interfaces are distinguished from those mentioned previously by the absence of additional liquid. The continuity of the electric field then becomes crucial and is thus ensured by different connection systems directly with the CE capillary. The objective is to obtain the most direct CE-ESI-MS interfacing in order to maximize the separation performance of CE as well as the MS sensitivity. Indeed, capillary diameters generally used in CE (20–75 μm) allow to provide EOF flow rate with an order of a few tens of nL/min. Such flow rates are similar to the conditions used in nano-ESI and promote ionization efficiency [38]. This explains the superior performance of sheathless interfaces compared with their counterparts, which use additional liquid [39,40]. The first CE-ESI-MS interface, introduced by Olivares et al. in 1987, was in fact a sheathless interface [11]. That design was later abandoned for several years in favor of sheath-liquid interfaces owing to their better compatibility and robustness with hyphenated techniques. However, sheathless interfaces have renewed interests over the last decade, not only because they meet the need for sensitivity that researchers now require but also because of a better control of the implementation processes. Several approaches have been developed to ensure the maintenance of the electrical contact, such as a conductive coating applied to the capillary tip, a wire inserted at tip or a porous etched capillary wall [4].

Certainly, the most widely developed method for achieving electrical contact is the coating of the capillary tip outer surface with a conductive material such as gold and silver. This type of interface is relatively simple to set up and allows to obtain higher sensitivities than that of sheath-liquid interfaces. Unfortunately, the major drawback of these interfaces arises from the lifetime of the coated tip. Repetitions of high-electrical field separations lead to premature wear of the conductive coating that subsequently leads to a rapid loss of the electrical contact.

Another way to perform sheathless CE-ESI-MS consisted of inserting a wire electrode into the capillary channel in order to maintain the electrical field. Different developments of this strategy have been described, such as the direct insertion of the wire electrode in the capillary tip when larger-inner-diameter capillaries are used [41] or into a small hole pierced near the end of the capillary [42]. In general, this interface system has similar characteristics to other sheathless interfaces. However, the presence of the wire inside the capillary tends to create turbulences that involve EOF disruption and may have negative consequences on the separation and coupling performance (decrease of resolution and alteration of the spray).

An alternative to produce more robust sheathless CE-ESI-MS interfaces has been developed by the group of Moini. It consists of manufacturing a porous section of the outside surface of the fused silica capillary by hydrofluoric acid (HF) etching. By this process, the porosity of the surface allows electrons and small ions to pass through the capillary walls. The electrical field can thereby be maintained through the capillary wall by inserting the porous section in a stainless-steel cannula filled by a conductive liquid. Since 1997, Moini's group developed different strategies to increase the robustness of their sheathless interface [42,43]. In 2007, they developed a simplified and more robust version by using a separation capillary having the outlet etched with HF to provide a 4- to 5-cm outer porous terminus [44] (Figure 5.1d). In 2014, Sciex Separation (Brea, USA) marketed this interface named CESI 8000. Performances of this interface have been assessed in a large number of applications such as proteomic [45], metabolomic [46], and intact protein analysis [47]. These works underline the robustness of the generated spray and the significant reduction of LODs and quantification compared with sheath liquid interfaces [14,39]. Indeed, particularly for the application of protein separation, the group of Somsen reported a comparison study of LODs for four model protein by using commercialized sheath liquid interface and CESI8000 device. They demonstrated that detection limits were improved by a factor of 6.5–20 with sheathless CE-MS [14].

#### 5.2.1.2 CE-MALDI-MS and 2D-CE-ESI-MS Interfaces

##### *5.2.1.2.1 CE-MALDI-MS Interfaces*

An attractive alternative to CE-ESI-MS is proposed with the coupling between CE and MALDI-MS. Indeed, several advantages of this hyphenation, such as a better tolerance to salts, sample collection with possibility of individual treatment (enrichment and digestion), and independent optimization of separation and detection condition, explain the strong and recent interest in this coupling. The first strategy to hyphenate CE with MALDI-MS consisted of developing online interfaces [5]. However, because sample is first dried on a solid surface before insertion into the MALDI source, heavy instrumental development appeared as one of the most significant issues for the implementation of this interface. This strategy was abandoned in favor of offline interfaces. As both ESI-MS and MALDI-MS are post-capillary detections, the major issue for the offline CE-MALDI-MS coupling is to maintain the electrical field during the separation and the fraction collection process. Several reports on instrumentation developments were published in the last decades [3,7]. Two approaches based on several capillary outlet configurations used in CE-ESI-MS have been adapted for CE-MALDI-MS: a sheath-liquid interface and a sheathless interface. In 2009, Busnel et al. designed an iontophoretic fraction collection approach, using a silver coating applied to the capillary tip to avoid the sheath flow (Figure 5.1e) [33]. On the same system, Gasilova et al. analyzed allergenic milk proteins by immuno-affinity CE-MALDI-MS [48]. However, this strategy suffers the same drawback described previously for CE-ESI-MS interfaces, namely the lifetime of the coated tip. For this reason, the majority of the existing interfaces is based on the coaxial sheath-flow interface using a T-junction. Recently, the Leize-Wagner's group described a fully automated offline CE-MALDI-MS interface with additional ultraviolet (UV) detection and integrated delivery matrix system [49]. Using this interface, authors performed the first analysis of intact monoclonal antibody (mAb) charge variants by capillary zone electrophoresis (CZE), using a MS detection. To date, no interface has been marketed, certainly owing to technical constraints.

#### *5.2.1.2.2 2D-CE-ESI-MS Interfaces*

Today, the instrumental developments regarding CE-MS are not only confined to interfacing efficiently both techniques. In a recent work, Neusüß's group described a system allowing two-dimensional (2D) CE-ESI-MS [50]. This instrumental development uses a set of mechanical valves to perform two consecutive electrophoretic separations in order to enhance electrophoretic resolution and peak capacity (Figure 5.1f). The outlet of the second capillary was positioned inside the ESI source by the intermediate of a regular sheath-liquid CE-MS interface. Regardless of the position of each valves, the CE capillaries extremities are always connected to the electrodes, which prevents variations, shortage of the electrical fields, or current leakages. Despite the apparent complexity of the system, all elements (valves and connectors) are commercially available and then can be assembled at home. As incompatibility of the CE separation medium components is a major limitation in hyphenation of CE to MS, 2D CE-ESI-MS could be used to realize in the first dimension a CE separation (CZE and capillary isoelectric focusing [CIEF]), using ESI-MS interfering BGE. To enable its hyphenation to ESI-MS, CZE using a volatile BGE was performed in the second dimension [34,51]. Neusüß's group reported the evaluation of this interface for biopharmaceutics.

## 5.2.2 Practical Issues of CE-MS Coupling

### 5.2.2.1 CE Modes and Coupling Strategies

For the analysis of intact proteins, three CE modes are largely used: CZE, CIEF, and capillary gel electrophoresis (CGE). Up to now, CZE has been the most frequently used mode for CE-MS analyses of proteins [15–17,52]. Indeed, CZE is characterized by its great simplicity of implementation, since the capillary is merely filled with a BGE and analytes are separated based on differences in their electrophoretic mobility ($\mu_{ep}$). Using sophisticated electrolytes (nonvolatile components) at often high concentrations, high separation efficiencies (theoretical plates up to one million or more) can be obtained, as longitudinal diffusion is ideally the only source of band broadening. Simple and MS compatible (i.e., volatile) BGE can be employed, such as formic/acetic acid and ammonium acetate/bicarbonate buffers. In addition, by selecting a physiological pH, proteins can be studied in their native state. The major drawbacks are linked to the limited number of volatile electrolytes compatible with MS and the need of low buffer concentrations. Indeed, a too high quantity of salts entering the MS detector may cause ionization suppression, source contamination, and high background signals [16]. Another drawback of CZE-MS is linked to the relatively poor LODs obtained with CZE, owing to the limited amount of sample that can be introduced into the capillary. Therefore, several strategies have been developed to preconcentrate the samples before the CZE analysis, in order to obtain high-sensitivity MS detection (see Section 5.2.2.3).

CIEF is also frequently used in CE-MS. It is a powerful strategy to separate, in the presence of ampholytes, amphoteric molecules according to their isoelectric point (p*I*) into narrow zones along a pH gradient created through an electric field (*E*). The molecules migrate through *E*, as long as they possess a charge. Reaching the position where the pH equals their p*I*, the net charge becomes zero, and they get focused in this region. The most common kind of CIEF is achieved by applying a voltage to an ampholyte system inside a neutral coated capillary to create a continuous pH gradient and no EOF. The whole process of CIEF can be divided into two steps: first, the focusing step with an acid solution at the anode end and an alkaline solution at the cathode end; and, second, the mobilization step, during which the analytes are pushed toward the detection system (by applying a pressure to the capillary inlet or by modifying the electrolyte composition of one reservoirs to achieve a chemical mobilization). An advantage of CIEF is that very low LODs can be reached. Indeed, not only can the entire capillary be filled with sample, but also the focusing of proteins induces a preconcentration effect, resulting in a higher peak intensity. One disadvantage of CIEF is that the mobilization step causes either focused band broadening (hydrodynamic mobilization) or distortion of pH gradient (chemical mobilization). However, this can be minimized by applying an electrical field during mobilization. Imaging CIEF (iCIEF) is another alternative, since it abolishes the mobilization step, the IEF process within the entire capillary being monitored online by the whole-column detection system.

With CIEF, ionization suppression effects are much more pronounced than with CZE-ESI-MS, owing to the presence of ampholytes, acids, and bases in the anolytes and catholytes. Different approaches have been developed to couple CIEF with MS [51,53]. CIEF-MS with catholyte/spray solution exchange after focusing is by far the most frequently used technique. It consists of placing the catholyte vial inside the ESI housing for focusing, switching to the ESI sprayer and sheath liquid for mobilization and detection [54]. This technique enables fast and reproducible CIEF-ESI-MS analyses and can achieve LODs that are two orders of magnitude smaller than those achieved with CZE for standard proteins. However, it requires one capillary-replacing step and a voltage interruption between focusing and mobilization. This may result in a loss of separation efficiency and hinders automation. To avoid the negative effect of ampholytes, acids, and bases on the ESI process, a CIEF-MS approach consists of using a membrane or a dialysis device integrated between the separation capillary and the ESI source. Another way to avoid the entry of interfering compounds into the ESI-MS is the placement of a free-flow electrophoresis (FFE) unit between the CIEF separation and the MS detector [55]. This setup enables to fully separate the analytes from interfering molecules but results in poor separation efficiency. Another strategy consists of coupling an liquid chromatography (LC) system between the CIEF and the ESI-MS instrument. However, this often requires complex and expensive interfaces and results in longer run times.

CIEF can also be coupled to MALDI-MS (see Section 5.2.1). Like with CIEF-ESI-MS, compounds such as ampholytes and catholytes are responsible for the decrease of MS signal intensities (e.g., the use of 1% ampholyte induces a signal decrease of 50%) [53]. On the other hand, it might be argued that the co-crystallization process in MALDI is a purification process itself, which might help in avoiding unfavorable effects of ampholytes on ionization yield. However, this has not been sufficiently documented in the reported offline CIEF-MALDI-MS studies. The first interface for offline coupling of CIEF to MALDI-time-of-flight (TOF)-MS, using a Tee junction, was reported by Karger's team [56]. Since then, several designs have been proposed [57–60]. Yet, all these concepts had to deal with interference from ampholytes, limited resolution and sensitivity for larger analytes, and limited separation efficiency under more MS-friendly conditions.

For the analysis of intact proteins, CGE is another option. CGE is particularly well suited to the analysis of dimers or oligomers of proteins. However, this mode is less suitable to CE-MS analysis of proteins, owing to the presence of nonvolatile separation media (i.e., polymers and surfactants). The strategy developed by Neusüβ's group to remove sodium dodecyl sulfate (SDS) from denatured protein samples by co-injection of cationic surfactant prior to MS analysis [61] could be applied to CGE-MS analysis. However, further investigations are required. At the present time, CGE remains a difficult option for CE-MS analysis of intact proteins.

#### 5.2.2.2 Capillary Coatings

A major issue in CE is the adsorption of proteins onto the fused-silica capillary wall (owing mainly to electrostatic or hydrophobic interactions), which degrades the separation performance. However, adsorption can be minimized through different strategies, the most efficient being the coating of the capillary. This can be accomplished via static (permanent or semi-permanent) coatings. For CE-MS of intact proteins, various capillary coatings have been proposed. Coatings of the silica capillary have a strong impact on the generated EOF, which in turn can lead to different flow rates at the exit of the capillary and have an impact on the spray stability. In addition, leakage of coatings is a major concern when dealing with CE-MS.

Positively charged coatings such as polybrene (PB) [62] and polyethyleneimine (PEI) [63] proved to be useful for CE-MS analysis of proteins. Under these separation conditions, both the proteins and the capillary wall are positively charged, avoiding protein-wall interactions and providing highly efficient separations. These charged coatings generate an appreciable and constant EOF toward the capillary outlet when a reversed polarity is applied, enabling adequate and reproducible CE-MS interfacing. To improve the stability, the reproducibility, and the lifetime of physisorbed capillaries, Katayama et al. [64] introduced, in 1998, successive multiple ionic-polymer layer (SMIL) coatings, the principle of which consists of sandwiching a cationic polymer between an anionic polymer and the negatively charged surface of the capillary by noncovalent bonding (Figure 5.2). Katayama's group developed a double-layer coating PB-dextran sulfate (DS), with a negative surface, for the analysis of acidic (p$I$ < 7) proteins at neutral pH [64], and a triple-layer coating PB-DS-PB, positively charged, for the analysis of basic (p$I$ > 7) proteins at low pH [65]. A few years later, Somsen's team largely exploited PB-polyvinyl sulfonic acid (PVS) [66–68] and PB-DS-PB [67,69,70] multilayer coatings for the analysis of intact proteins. In parallel to SMIL coatings, positively charged copolymers such as

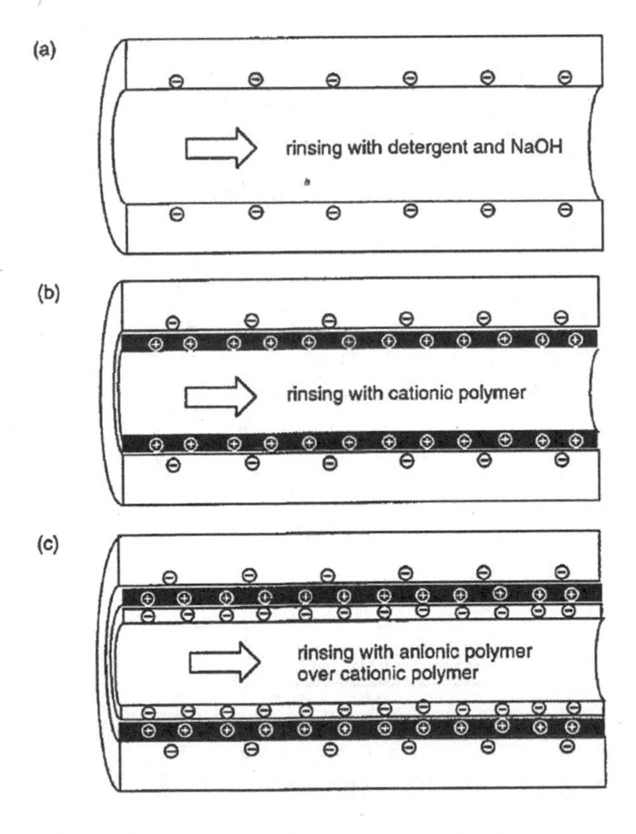


**FIGURE 5.2** SMIL coating procedure. (a) Activation of the silanol groups. (b) First layer coating. (c) Second layer coating. (Reprinted from Katayama, H. et al., *Anal. Chem.*, 70, 2254–2260, 1998. Copyright 1998 American Chemical Society. With permission.)

*N,N*-dimethylacrylamide-ethylpyrrolidone methacrylate (DMA-EpyM) have been developed by Cifuentes' group [71]. DMA-EpyM has been successfully applied to the CE-MS analysis of intact proteins [63,72,73].

The disadvantage of the high EOF generated by these positively or negatively charged capillaries is that the analyte resolution can be limited. For CE-MS of intact proteins, neutral coatings (EOF near zero) such as hydroxypropylcellulose (HPC) [74–76], polyvinyl alcohol (PVA) [61,77], and polyacrylamide (PAA) [61–63] have also been investigated. Taichrib et al. [63] compared the use of several neutral coatings (dynamic PAA [UltraTrol™ LN], permanent PAA, and Guarant™ composed of thermally immobilized galactomannans guaran and locust bean gum) and positively charged coatings (PEI, cationic PAA [UltraTrol™ HR], and DMA-EpyM copolymer) for the analysis of model proteins as well as erythropoietin (EPO) with sheath-liquid CE-MS, using low-pH BGE. The neutral coatings all resulted in higher separation resolution when compared with the positively charged coatings, at the expense of longer analysis times. Moreover, it was demonstrated that the nebulizer gas, commonly used in sheath-liquid interfacing, significantly increased the residual EOF of the neutral coated capillaries, owing to suction. This shortened the protein's migration time and could have a negative effect on the separation efficiency if the suction phenomenon was too strong.

#### 5.2.2.3 Preconcentration Techniques

Over the past decades, different preconcentration techniques have been developed to increase the sensitivity of CZE-MS analyses of intact proteins. They can be classified into two categories: those that are performed by electrokinetic processes and those that use a solid support present at the inlet of the separation capillary.

Among the electrokinetic approaches, large-volume sample stacking (LVSS) has been applied in CZE-MS analysis of proteins to enhance the loading capacity [78,79]. In LVSS of acidic proteins, a large plug of low-conductivity sample with a pH above the *pI* of the proteins is hydrodynamically injected in a fused-silica capillary, previously filled with a BGE of relatively high conductivity. Then, a negative voltage is applied, and the EOF leads the large plug in the direction of the capillary inlet, hence pushing this plug out of the capillary. Meanwhile, the negatively charged proteins are subjected to a strong local electric field and consequently move with a high electrophoretic velocity toward the boundary between the sample plug and the BGE zone. Once the proteins reach this boundary, their velocity is slowed down in the lower electric field of the higher conductivity buffer, resulting in the stacking of the analytes. When the main part of the sample plug is removed and most of the proteins are concentrated in a narrow zone, the voltage polarity is switched to positive, and the separation starts under normal CE conditions. For basic proteins (positively charged at pH $< pI$), the same procedure can be used. In this case, a positive stacking voltage should be applied, and a positively charged coated capillary should be employed in order to reverse the EOF. The concentration factors typically obtained with LVSS in CE-MS are around 10.

Another way to increase the loadability of proteins before the CZE separation is to perform an isotachophoresis (ITP) preconcentration step [80,81]. The ITP step takes place in a discontinuous electrolyte system consisting of a leading electrolyte of high mobility and a terminating electrolyte of low mobility. Proteins, which are introduced in a relatively large plug, with mobilities between the two electrolytes, will be focused during the CZE separation step to form narrow zones. For CE-MS of proteins, the most commonly used leading and terminating electrolytes are ammonia and acetic acid, respectively. Concentration factors between 10 and 50 can be obtained, using injection plugs between 15% and 45% of the capillary volume.

Solid-phase extraction (SPE) is another way to achieve the preconcentration of proteins in CZE [82,83]. The micro-SPE cartridge, filled with, for example, C18 or C8 reversed-phase material, is inserted at the inlet side of the CE capillary, allowing the loading of a relatively large volume of sample. Proteins are trapped on the SPE material, and matrix components are removed via a rinsing step. After preconcentration, the proteins are desorbed in a small volume into the CE capillary and separated by CZE. Recently, Sanz-Nebot's group developed an online immunoaffinity (IA)-SPE-CE-MS method using magnetic beads (MBs) for the analysis of serum transthyretin (TTR), a homotetrameric protein (*Mr* ~ 56,000) involved in different types of amyloidosis [84]. The suitability of three protein A (ProA) MBs (ProA Ultrarapid Agarose™ [UAPA], Dynabeads® ProA [DyPA], and SiMAG-ProA [SiPA]), and AffiAmino Ultrarapid Agarose™ (UAAF) MBs to prepare an IA sorbent with an intact polyclonal antibody (Ab) against TTR was studied. Figure 5.3 presents the

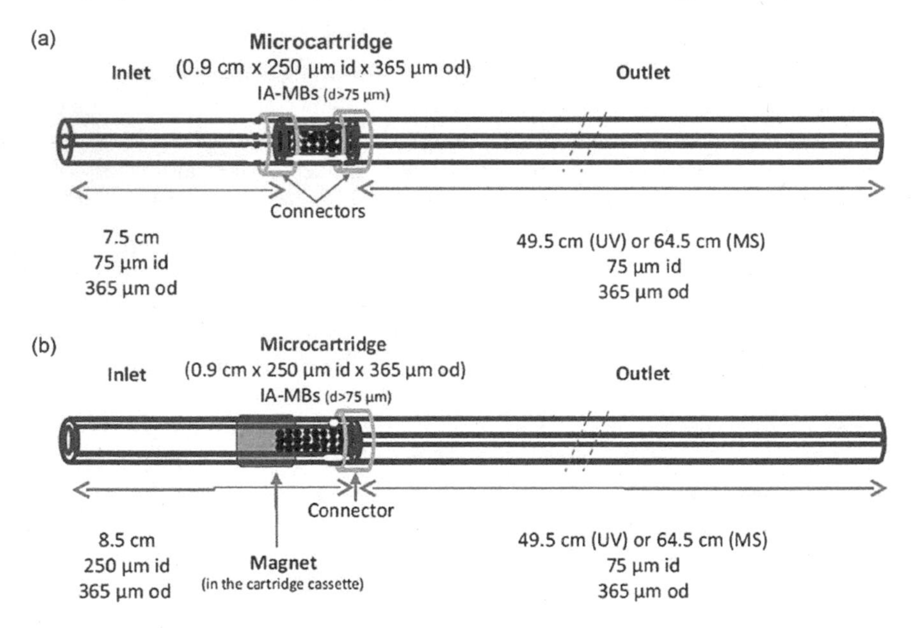


**FIGURE 5.3** Representations of the microcartridge designs (a) UAPA or UAAF MBs are trapped in a microcartridge body of 250 µm internal diameter (id), owing to their particle size, and (b) UAPA or UAAF MBs are retained in one of the ends of a piece of 250 µm id capillary, and a magnet prevents the shift and loss of the MBs. (The first design could not be applied with SiPA and DyPA MBs, because both are very small. Similarly, in the second case, the magnet should cover the whole microcartridge body). (From Peró-Gascón, R. et al.: Analysis of serum transthyretin by on-line immunoaffinity solid-phase extraction capillary electrophoresis mass spectrometry using magnetic beads. *Electrophoresis*. 2016. 37. 1220–1231. Copyright Wiley-VCH Verlag GmbH & Co. KGaA. Reproduced with permission.)

microcartridge designs. All the different MBs enabled the identification and quantitation of the relative abundance of the six most abundant TTR proteoforms. For online immunopurification, UAAF MBs were selected, because Ab was not eluted from the MBs. The LODs achieved with this IA-SPE-CE-MS methodology, using TTR standards, were 25 times lower than those obtained by CE-MS (~1 µg/mL and ~25 µg/mL, respectively). Using Fab′ antibody fragments covalently attached to succinimidyl silica particles, the same group managed to decrease the LODs slightly (2-folds) for the analysis of TTR, which allowed the detection of proteoforms found at lower concentrations [85].

## 5.3 FROM DIMER TO OLIGOMER ANALYSIS OF PROTEINS

Proteins derived from plasma as well as recombinant ones are subjected to the modification of their spatial structure, owing to bioprocessing steps (pasteurization, ultra-/diafiltration, and lyophilization) or during protein handling and storage, leading to their denaturation and formation of oligomers or aggregates. Many techniques, including X-ray crystallography, nuclear magnetic resonance, high-performance size-exclusion chromatography, native polyacrylamide gel electrophoresis, analytical ultracentrifugation, CE-UV, MS, and ion-mobility spectrometry coupled to MS, are available for studying protein oligomerization. CE-ESI-MS is an alternative technique to analyze protein complexes, but the challenge consists of maintaining weak assemblies. Up to now, sheath-liquid interfaces have been used in most of the reported studies to analyze dimers or oligomers of proteins but using different CE modes, including CZE, ACE, and CIEF. Although the use of CE-ESI-MS is becoming more popular for complex protein analysis these last decade, there are not many studies describing the major factors impacting the performance of this method for noncovalent protein complexes. Indeed, special attention should be paid on separation and detection conditions not to cause protein unfolding or disrupt weak forces holding noncovalent complexes, oligomers, or aggregates. The most important challenge is to select (i) nondenaturing electrophoretic separation conditions but providing enough high-resolution power, (ii) sheath-liquid containing optimal organic solvent and acid quantity, allowing the optimal protein detection sensitivity, and (iii) soft MS parameters to prevent in-source dimer/oligomer dissociation or association but sufficiently efficient in terms of ionization.

In 1999, Gysler et al. reported a CZE-MS method for monitoring the recombinant human interleukin-6 (rhIL-6) dimer formation throughout an acidic incubation, using a 20 mM ammonium acetate buffer, pH 4.2, and a sheath-liquid consisting of methanol-water (80:20, v/v) with 1%(v/v) acetic acid [80]. Unfortunately, they failed to detect the dimer (detection of only the monomeric rhIL-6) in MS, while the CZE-UV method showed two peaks for the dimer, supporting that rhIL-6 forms two different dimeric complexes. This finding also indicated the dissociation of the noncovalently bound dimeric rhIL-6 complexes during the mixing with the methanol-containing sheath-liquid (methanol in high quantity). This concept was supported by complete disappearance of the dimer peaks and the simultaneous increase of the monomer peak when 30% methanol was added to samples containing dimeric rhIL-6. Other authors have also demonstrated that the composition of the sheath-liquid greatly affects the dissociation of intact noncovalent protein complexes [86,87].

Beside this parameter, Sanz-Nebot's group showed that in the case of the bovine Cu-Zn superoxide dismutase (SOD-1), the acidity of the BGE was the main factor influencing the detection of the native form of SOD-1 [88]. The native form is a homodimer that coordinates one $Cu^{2+}$ and one $Zn^{2+}$ ions per monomer; these ions play crucial roles in enzyme activity and structural stability, respectively. The dimer formation is essential for SOD-1 functionality, and in humans, several SOD-1 mutant forms have been associated to certain types of amyotrophic lateral sclerosis. Using 1 M acetic acid BGE (pH 2.3), the analysis of the SOD-1 dimer showed a separation into one minor and one broad peak. The calculated molecular mass (15601 Da) after deconvolution of the major peak was close to the theoretical molecular mass of the metal-ions-free monomeric SOD-1 (apo-SOD-1) (Figure 5.4a), suggesting that acidic electrophoretic conditions generates apo-SOD-1, for example, dimer dissociation and complete release of metal ions. Using 10 mM ammonium acetate BGE (pH 7.3), the mass spectrum of the main electrophoretic peak showed the presence of the monomer (Cu-Zn SOD-1) and the dimer ($Cu_2$-$Zn_2$ SOD-1), with a molecular

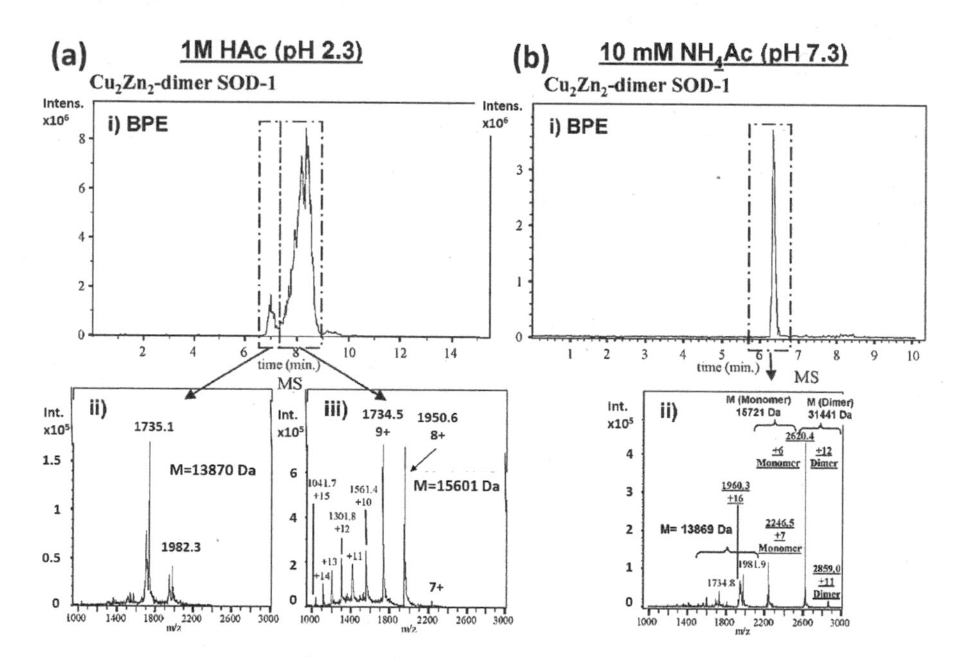


**FIGURE 5.4** CZE-ESI-MS base peak electropherogram (BPEs) for (i) $Cu_2$-$Zn_2$ SOD-1 samples using a BGE of (a) 1M HAc (pH 2.3) and (b) 10 mM $NH_4Ac$ (pH 7.3). The molecular mass values obtained after deconvolution are shown in the insets of the mass spectra (ii) and (iii). Conditions: MS-positive mode; sheath-liquid, isopropanol-water 60:40 (v/v) with 0.5% formic acid delivered at a flow rate of 0.2 mL/h; CE voltage, 15 kV; injection, 10 s at 50 mbars, fused silica capillary. (From Borges-Alvarez, M. et al.: Capillary electrophoresis/mass spectrometry for the separation and characterization of bovine Cu, Zn-superoxide dismutase. *Rapid Communications in Mass Spectrometry*. 2010. 24. 1411–1418. Copyright Wiley-VCH Verlag GmbH & Co. KGaA. Reproduced with permission.)

mass of 15721 Da and 31441 Da, respectively (Figure 5.4b). The authors verified that the acidic sheath-liquid containing isopropanol/water (60/40: v/v) and 0.1% formic acid did not contribute to the enzyme dissociation but affected only the sensitivity, by performing experiments without formic acid and obtaining similar results. This suggests that the monomer was already present in the original dimer solution and was not an artifact of ionization. The authors also claimed that the neutral BGE provided enhanced detection of the native enzyme (dimer form), taking into account that MS parameters are finely tuned to prevent the disruption of noncovalent protein complexes from solution into gas phase.

In a second paper, these authors also highlighted the importance of selecting the appropriate fragmentor voltage to minimize in-source induced collision dissociation of the human SOD-1 dimer to obtain reliable quantitative information [89]. When changing the fragmentor voltage from 275 V to 375 V, the proportion of the monomer relative to the dimer increased, showing in-source fragmentation (Figure 5.5).

The first experiments on bovine SOD-1 were performed using an ion-trap analyzer that enabled to scan up to 3000 *m/z*; this is the reason why it allowed to detect only the ion with +11 charge state of the dimer SOD-1 [88]. In the second paper studying human SOD-1, a TOF mass spectrometer was used, and the scanning range was extended until 4500 *m/z*, allowing the detection of the ion with +10 charge state of the dimer SOD-1 [89]. This shows that the TOF analyzer, and in a wider manner the QTOF, is totally adapted for the analysis of multimer proteins when the cluster of multicharged ions of the proteins is displaced to high *m/z* values.

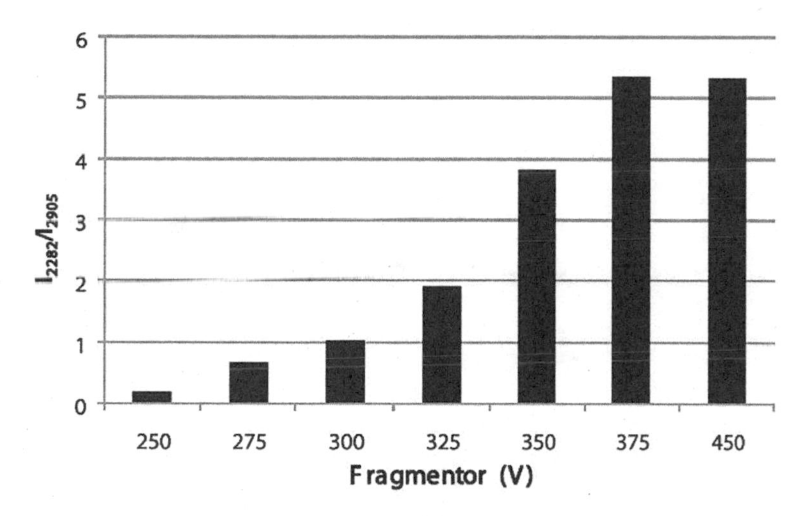


**FIGURE 5.5** Bar graph of the ratio between the intensity of the ions corresponding to monomer and dimer of human SOD-1 at different fragmentor voltage values. Conditions: Fused silica capillary BGE; 10 mM $NH_4Ac$ (pH 7.3); MS-positive mode; sheath-liquid, isopropanol-water 60:40 (v/v) with 0.5% formic acid delivered at a flow rate of 3.3 μL/min; CE voltage, 15 kV; injection, 10 second at 50 mbar. (From Borges-Alvarez, M. et al.: Separation and characterization of superoxide dismutase 1 (SOD-1) from human erythrocytes by capillary electrophoresis time-of-flight mass spectrometry. *Electrophoresis*. 2012. 33. 2561–2569. Copyright Wiley-VCH Verlag GmbH & Co. KGaA. Reproduced with permission.)

In 2015, the same group had described a method for the detection and characterization of normal and variant forms of TTR in serum samples from healthy controls and familial amyloidotic polyneuropathy type I (FAP-I) patients. FAP-I is associated with a mutant TTR form (substitution of Val for Met30), the native TTR form being a tetrameric structure [90]. In contrast to previous studies on homodimeric SOD-1 of this group [88,89], the neutral BGE, which is generally advantageous for detecting the native form of protein, allowed to detect only the monomer and dimer but not the native tetramer in the case of TTR. The authors attributed this to the tetramer disruption, either through the acidic sheath-liquid, as 0.25% of formic acid v/v in the sheath-liquid was necessary to detect the protein, or through the high vacuum-pressure inside the mass spectrometer. Furthermore, the TTR concentration in serum samples is too low, leading to the use of acidic BGE that produces higher sensitivity. Finally, this study did not require working under nondenaturing conditions, as detecting the monomer TTR (nonnative form) was sufficient to distinguish control individuals from FAP-I patients.

Nguyen and Moini developed a CZE-ESI-QTOF-MS to detect a protein complex in red blood cells, hemoglobin (Hb), that exists as a tetramer (average MW 66446) consisting of four noncovalently bonded protein subunits (two α chains and two β chains, with each chain attached to a heme group) [91]. In addition to the BGE pH (in this case physiological pH), which is one of the critical factors, the polarity of the CE separation was also found important for detecting Hb tetramer. Surprisingly, under positive ESI mode, the tetramer detection was made possible only under normal (called forward)-polarity CE, while Hb dimer was observed rather under reverse-polarity CE (Figure 5.6).

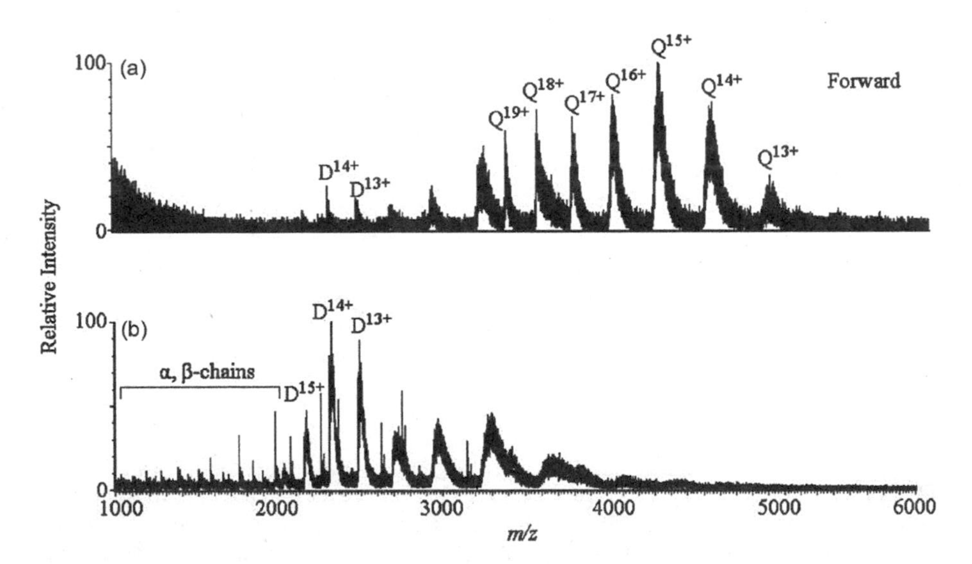


**FIGURE 5.6** Mass spectra of the Hb in (a) normal-polarity CE and (b) reversed-polarity CE. The Hb-dimer and tetramer are indicated by D and Q, respectively. Conditions: Polybren-coated capillary BGE; 1.7 mM $NH_4Ac$ (pH 7.4); CE voltage: +/−30 kV; MS-positive mode; sheathless interface. (Reprinted with permission from Nguyen, A. and Moini, M., *Anal. Chem.* 80, 7169–7173, 2008. Copyright 2008 American Chemical Society.)

The authors explained the complete dissociation of the tetramer in reverse-polarity mode and minor fragmentation in normal-polarity mode by two electrochemical reactions occurring simultaneously at the CE outlet/MS inlet electrode. Under reverse-polarity CE, the shared electrode acts as the anode for both the CE and ESI electrical circuits, leading to an important decrease of the solution pH (from 7.4 to ≅5) that dissociated all the tetramer. In this case, all dissociation products comigrate in a single electrophoretic peak, as the dissociation is happening close to the CE capillary outlet.

However, until 2016, CE has never been hyphenated to MS for monitoring protein dimer formation and detecting the dimeric state. Francois et al. performed the characterization of mAbs $F_{c/2}$ dimers by offline CZE-UV/ESI-MS. Authors reported an offline strategy to enable CZE-ESI-MS using CZE-UV/fraction collection technology, followed by ESI-MS infusion of the different fractions, using sheathless CE-ESI-MS interface as the nano-ESI infusion platform. This strategy allowed to highlight the presence of homo- and heterodimers of cetuximab $F_{c/2}$ fragments formed by possible lysine truncation [75]. In the same year, Marie et al. developed a CZE-ESI-QTOF-MS method using a sheath-liquid interface to analyze and separate native antithrombin (AT) and its dimer, which is an inactive form that decreases the quality and efficacy of therapeutic AT preparations administrated for their anticoagulant activity [77]. For the first time, the detection of the dimeric state of AT was observed. The deconvolution MS spectra yielded an average molecular mass of 115 kDa, which is in agreement with the expected mass of an AT dimer (homo- or heterodimer). Dedicated CE-MS experiments consisting of successive injections of native and latent forms (for additional information on conformer separation, see Section 5.4) separated by a small BGE plug present two partly resolved peaks. The mass spectrum of the later migrating peak corresponded to the native form, but the first migrating peak contained both latent and dimeric forms, suggesting that native and latent forms associate during the CE analysis to form a heterodimer. Thus, unfortunately, the method does not allow an accurate quantification of the AT forms in pharmaceutical AT preparations.

More recently, Ivanov's group reported the first attempt of the intact mAb characterization by using a native CZE-MS with unaltered noncovalent interactions and sheathless instrumentation [92]. High-efficiency separation of Trastuzumab components was shown by using a neutral PAA-coated capillary with HF-etched emitter and a BGE of 20 mM ammonium acetate, pH 8.0. The mass spectrum corresponding to the main electrophoretic peak revealed the presence of the monomer and less than 1% of dimeric species, while dimers of Trastuzumab could not be observed in denaturing CZE-ESI-MS experiments [93]. Deconvolution of this native CZE-MS in a selected target mass range and *m/z* range, shows intensities of the dimer peaks 140–250-folds lower than the intensity of the predominant monomer peak. This demonstrated that relative quantitation can be achieved by using this approach to evaluate the determination of mAb aggregation.

Daniel's group developed the coupling with MS to other CE modes that are more delicate to hyphenate than CZE because of the presence of several agents in the electrolyte. Fermas et al. showed the effect of glycosaminoglycans (GAGs) on the dimerization of chemokine stromal cell-derived factor-1 (SDF-1), using affinity capillary electrophoresis (ACE) hyphenated to MS [94]. A major effect of the GAG binding to SDF-1 may be on the quaternary structure of this protein, which has been reported as a dimer; this dimerization affects the presentation of SDF-1 to the receptor. The presence of fucoidan pentasaccharide in the BGE led to the detection of three charge state distributions in MS, corresponding to the monomeric SDF-1, the 1/1 SDF-1/fucoidan, and the 2/1 SDF-1/fucoidan complex. This indicated that the binding of this GAG promoted the SFD-1 dimerization. Przybylski et al. reported a CIEF-ESI-MS method using a glycerol/water medium in silica capillary instead of an aqueous gel medium in coated capillary, in order to avoid the contamination of the mass spectrometer [95]. Under positive ionization mode and using pH 3–10 ampholytes at 1.5% in glycerol/water 60/40 medium, interferon (IFN)-γ, which is a homodimeric cytokine, could be detected as its dimeric form in the presence of a volatile aqueous sheath-liquid composed of 10 mM ammonium acetate, pH 5.0. This indicates that IFN-γ migrates in CIEF under a dimer state and that the acetate buffer sheath-liquid enabled to keep the noncovalent association of IFN-γ monomers at the capillary exit in the ion source and up to the MS detection.

## 5.4 INVESTIGATION OF CONFORMATION AND FOLDING OF PROTEINS

The biological activity of a given protein depends greatly on its conformation. Indeed, the protein needs to be in a correct folded state (the so-called native structure) to be active. It is therefore of paramount importance to develop analytical methods capable of detecting conformational protein isomers.

CE is very attractive to study protein conformation, since near-physiological conditions can be employed. This contrasts with chromatographic methods, for which solvents and extreme pH conditions needed to repel proteins from ionized silanols are likely to affect the conformation. In addition, the fact that CZE separations are driven by differences in the charge to hydrodynamic volume ratio of the analytes renders this separation step quite helpful in distinguishing protein conformers. In the past two decades, CE-based methods have been developed to study folding/unfolding or misfolding of proteins [3,96–99]. For example, protein isoforms corresponding to different disulfide bridge arrangements have been detected using CZE [97] or CGE [98].

The analysis of protein conformers with MS is a complicated task, since no mass difference is expected between two conformers, and if it is relatively easy to distinguish a folded from an unfolded form (the latter exhibiting a higher degree of ionization), it is quite a challenge to separate two conformers in a folded state. Therefore, CE-MS coupling represents a promising approach for the analysis of protein conformers—CE being capable of separating two conformers and MS providing accurate molecular mass. Very recently, several groups investigated the potential of CE-MS to separate and analyze conformers of intact proteins.

In 2016, Marie et al. developed a CZE-ESI-QTOF-MS method by using a sheath-liquid interface for the characterization of native and latent forms of AT [77]. It has been demonstrated that only the native form of AT is active. Indeed, in this form, the reactive center loop (RCL) is exposed as a target for coagulation proteases (Figure 5.7a). The presence of latent AT could thereby decrease the quality and

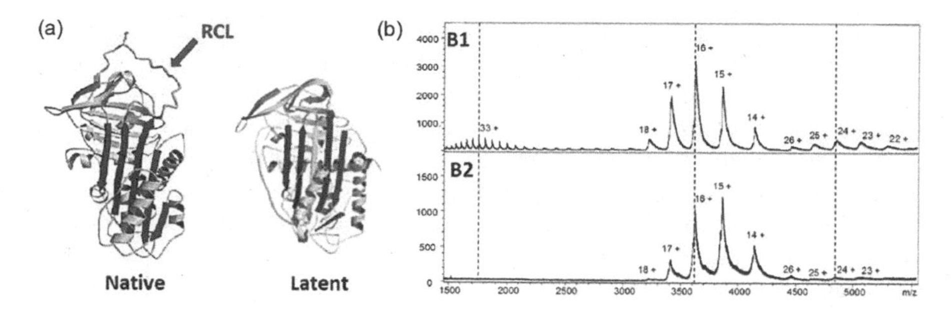


**FIGURE 5.7** (a) Native and latent forms of antithrombin. In the latent form, the reactive center loop (RCL) is buried inside the molecule, and no binding can occur with the coagulation proteases. (b) CE-MS analyses of AT (molecular mass, 57858 Da). Mass spectra obtained during the analysis of preparations containing a high proportion of (B1) AT native form (>70%) and (B2) AT latent form (>70%). Conditions: PVA-coated capillary; BGE, 50 mM $NH_4Ac$, pH 7.4; sheath-liquid, isopropanol-water 50:50 (v/v) with 14 mM $NH_4Ac$ delivered at a flow rate of 120 μL/h; CE voltage, −20 kV; capillary pressure during CE analysis, 0.7 psi; injection, 30 second at 0.5 psi. The charge state distributions centered at 33+ and from 18+ to 14+ correspond to monomeric forms of AT and those from 26+ to 22+ to dimeric forms of AT. (Reprinted from Marie, A.-L. et al., Characterization of conformers and dimers of antithrombin by capillary electrophoresis-quadrupole-time-of-flight mass spectrometry, *Anal. Chim. Acta*, 947, 58–65, Copyright 2016, with permission from Elsevier.)

efficacy of the drug. For CE separation, a PVA-coated capillary was employed. The preservation of the protein conformation was obtained using a BGE composed of 50 mM ammonium acetate, at pH 7.4. The sheath-liquid consisted of isopropanol-water 50:50 (v/v) with 14 mM ammonium acetate and was delivered at a flow rate of 120 μL/h. As shown in Figure 5.7b, the mass spectra of the two AT conformers presented differences regarding their respective charge state envelopes. For both conformers, a charge state envelope in the 3200–4200 *m/z* range was observed, but the charge state envelope of the native form was centered on the 16+ ion, whereas that of the latent form was centered at 15+ ion. In addition, for the latent form, no charge state envelope centered around 33+ ion in the 1500–2200 *m/z* region was observed in the mass spectrum, indicating that this conformer is more resistant to denaturation. The developed CZE-ESI-MS method is therefore capable to unambiguously distinguish the two AT conformers. The CZE-ESI-MS method was applied to a stability study. For this purpose, a commercial AT preparation was stored at room temperature for 3 weeks, before its analysis by CZE-ESI-MS. The mass spectrum obtained revealed the formation of a high quantity of latent form.

Additional information was obtained from the developed CZE-ESI-MS method. Especially, several co-injection experiments showed that the two AT conformers could associate to form a heterodimer (see Section 5.3) in the course of CE analysis. That is the reason why the developed method is not fully adapted to quality control of pharmaceutical AT preparations.

In the same year, Bertoletti et al. investigated different CE-ESI-MS setups for the analysis of conformational states of beta-2-microglobulin ($\beta_2$-m), a protein often taken as a model of amyloidogenic globular proteins [100]. For all experiments, a

BGE consisting of 50 mM ammonium bicarbonate, pH 7.4, and uncoated capillaries were employed. Two kinds of interfaces were tested using ESI ionization: a sheath-liquid interface coupled with a single quadrupole (SQ)-MS or a TOF-MS, and a sheathless interface coupled with high-resolution QTOF-MS. A sheath-liquid composed of isopropanol-water-formic acid (49.5:49.5:1, v/v/v) delivered at a flow rate of 4 μL/min allowed to obtain high ionization efficiency. $\beta_2$-m was analyzed either under native conditions (i.e., diluted in water) or in a partially denatured state after dilution in acetonitrile. By analyzing $\beta_2$-m with sheath-liquid CZE-ESI-SQ-MS, two peaks in dynamic equilibrium were separated, the ratio of which depended on the injected $\beta_2$-m sample (Figure 5.8a,b). The first peak in the total ion electropherogram was attributed to the native form, whereas the later migrating species to the partially unfolded conformer. Yet, no MS identification was possible, owing to the limited performance of the SQ. With sheath-liquid CZE-ESI-TOF-MS, different capillary lengths were tested (100, 200, and 300 cm). After injection of $\beta_2$-m partially denatured with acetonitrile, two peaks were observed, but the charge state distributions of each peak were clearly alike. Interestingly, the peak ratio appeared to depend on the capillary length (Figure 5.8c–e), certainly owing to protein refolding during CE analysis. The sheathless CE-ESI-QTOF-MS analysis of $\beta_2$-m after incubation with acetonitrile enabled the separation of three peaks (Figure 5.8f). As shown in Figure 5.8g, differences between the charge state distributions of peaks 1 and 2 were observed regarding the relative abundance of 6+ and 7+ charge states, indicating that peaks 1 and 2 corresponded to the native and partially unfolded forms of $\beta_2$-m, respectively. This result demonstrates the suitability of the sheathless interface to preserve the protein structure integrity.

Tengattini et al. reported a CE-MS method to monitor the integrity of an antigenic protein during glycoconjugate vaccine synthesis [70]. In order to prevent protein adsorption to the inner capillary wall and to achieve efficient separation of the antigen proteoforms, a BGE of 1.5 M acetic acid, pH 2.3, was selected, and the capillary was coated with PB-DS-PB. MS coupling was performed with a sheath-liquid interface and ESI ionization. The sheath-liquid consisted of isopropanol-water (50:50, *v/v*) with 0.1% formic acid and was delivered at a flow rate of 3 μL/min. The analysis of the protein Ag85B with the developed CE-MS method allowed to separate one main peak (0) from three relatively small ones (1–3) (Figure 5.9a). The mass spectra obtained for peaks 0 and 1 revealed, after deconvolution, a similar molecular mass of 31345 Da, indicating the separation of two Ag85B conformers. However, differences were observed in the charge state distribution. The mass spectrum of the main peak (Figure 5.9$a_1$) exhibited a shift of the distribution to lower *m/z* values, probably owing to partial unfolding of the protein during ESI. The mass spectrum of peak 1 (Figure 5.9$a_2$) showed a significantly lower relative abundance of the charge states at lower *m/z* values, indicating less susceptibility to unfolding. Deconvolution of the mass spectra of peaks 2 and 3 led to the same molecular mass of 30424 Da, corresponding to a truncated form of Ag85B. However, peaks 2 and 3 differed in their charge state distributions in a similar fashion as with peaks 0 and 1, again suggesting the presence of two protein conformers.

In parallel, in 2016, Berezovski's group [101] exploited an emerging technology called kinetic CE, hyphenated to UV detection and ion mobility MS (CE-UV-IM-MS),

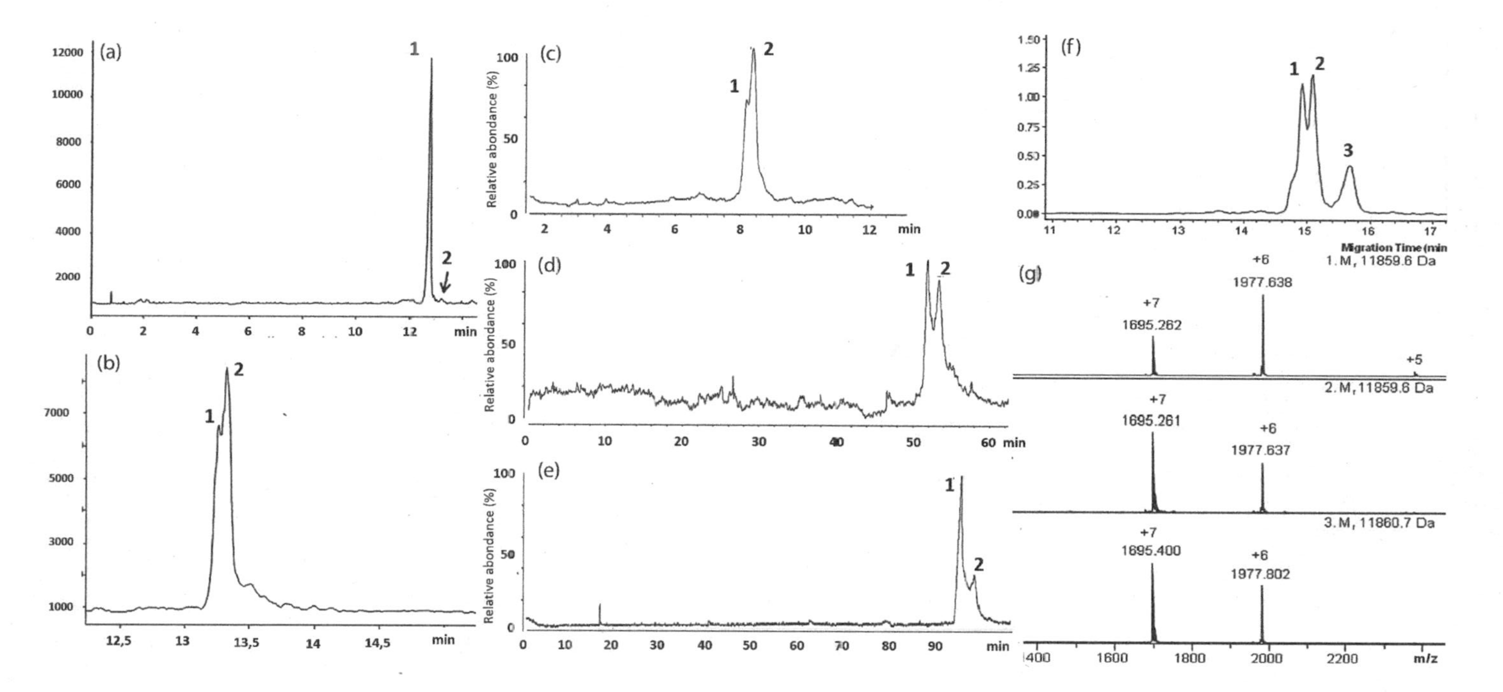


**FIGURE 5.8** CE-MS analyses of beta-2-microglobulin ($\beta_2$-m) (molecular mass, 11860 Da). Sheath liquid CE-ESI-SQ-MS analysis of $\beta_2$-m in (a) water, and (b) acetonitrile (extracted ion electropherograms). Sheath liquid CE-ESI-TOF-MS analysis of $\beta_2$-m in acetonitrile using a capillary length of (c) 100 cm, (d) 200 cm, and (e) 300 cm (total ion electropherograms). Sheathless CE-nano-ESI-Q-TOF-MS analysis of $\beta_2$-m in acetonitrile: (f) Cumulative extracted ion electropherograms of *m/z* 1695–1696 (6+) and 1977–1978 (7+), and (g) mass spectra of peaks 1–3. (Reprinted from Bertoletti, L. et al., Evaluation of capillary electrophoresis-mass spectrometry for the analysis of the conformational heterogeneity of intact proteins using beta2-microglobulin as model compound, *Anal. Chim. Acta*, 945, 102–109, Copyright 2016, with permission from Elsevier.)

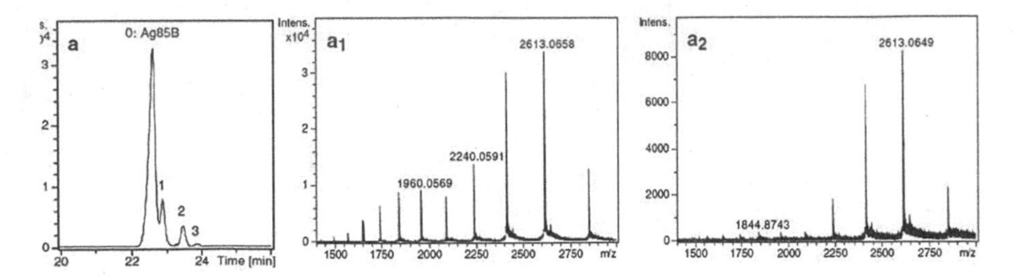


**FIGURE 5.9** CE-MS of Ag85B (molecular mass, 31345 Da). (a) Base peak electropherogram (BPE). ($a_1$) and ($a_2$) mass spectra obtained at the apex of peaks 0 and 1, respectively. (From Tengattini, S. et al. Monitoring antigenic protein integrity during glycoconjugate vaccine synthesis using capillary electrophoresis-mass spectrometry, *Anal. Bioanal. Chem.*, 408, 2016, 6123–6132. With permission from Springer Science+Business Media.)

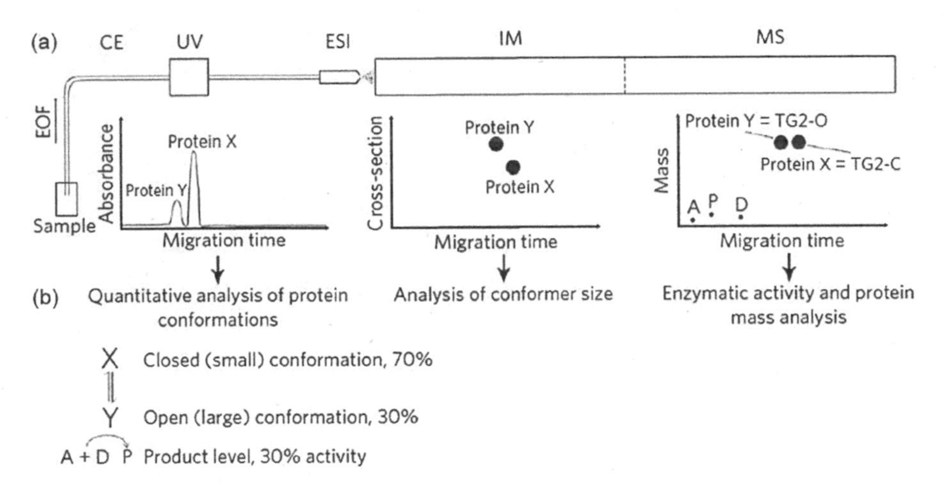


**FIGURE 5.10** Schematic representation of CE-UV-IM-MS method. (a) A, acceptor substrate; D, donor substrate; P, product; TG2-O, "open" conformation of TG2 (protein Y); and TG2-C, "closed" conformation of TG2 (protein X). (b) A theoretical model of conformation changes and enzymatic activity of TG2. Proteins X and Y, with relative abundance of 70% and 30%, respectively. (Mironov, G. G. et al. Simultaneous analysis of enzyme structure and activity by kinetic capillary electrophoresis–MS, *Nature Chem. Biol.*, 12, 2016, 918. With permission from Springer Science+Business Media.)

for simultaneous analysis of enzyme structure and activity (Figure 5.10). The human tissue transglutaminase (TG2, *Mr* ~78 kDa) was used as a model. In the crystal structure of the guanidine diphosphate (GDP)-bound form, TG2 adopts a compact, "closed" form that restricts access to its active site. In contrast, the crystal structure of TG2 following its calcium-dependent reaction with an irreversible inhibitor reveals an extended, "open" conformation. The biological activity of TG2 is thought to be regulated through a dramatic conformational change between the open and closed forms. Previously, the same group [102] had shown that the slow interconversion of these two conformers allowed them to be separated and monitored by kinetic CE (first introduced by Berezovski and Krylov in 2002 [103] and defined as the separation of molecules that interact inside a capillary during electrophoresis). By coupling online this CE method (using a BGE composed of 15 mM Tris-acetate, pH 7.85, and uncoated capillaries) with IM and MS (using ESI ionization), a simultaneous, real-time observation of the effect of small molecule inhibitors on both the conformational distribution and enzymatic activity of TG2 becomes possible. Indeed, kinetic CE-UV allows to separate protein conformers and monitor their interconversion dynamics in solution; IM estimates the size of each TG2 conformer; and MS informs on the mass of each conformer and quantifies the added substrates (donor and acceptor) and the product generated by the TG2-mediated transamidation reaction.

Up to now, CE-MS has been rarely exploited for studying protein folding and conformers. However, the recent works reported in this chapter prove that, with the advance of ESI-MS instruments and CE-MS interfaces, CE-MS coupling is an emerging technique in this field. Intact proteins in their most native-like state can be analyzed, and, in the future, new applications of CE-MS could arise, such as the analysis of noncovalent protein-protein or protein-ligand complexes.

## 5.5 MAPPING THE HETEROGENEITIES OF INTACT PROTEINS

### 5.5.1 Post-Translational Modifications

PTMs refer to the covalent modification of proteins during or after protein biosynthesis, generally by enzymes. PTMs contribute to the structural variety of proteins and can significantly affect their physiochemical properties and biological activity. There are different types of PTMs, such as phosphorylation, glycosylation, acetylation, and sulfonation. Characterization of PTMs is crucial and challenging; it involves the development of hyphenated techniques such as HPLC and CE coupled to MS. CE-MS is a powerful tool, thanks to its high sensitivity, speed, and peak capacity, allowing subtle structural variations such as positional isomers to be detected. Unlike liquid chromatography techniques, mild conditions without the need of solvents can be used, avoiding protein denaturation during analysis.

CE-MS separation of intact protein PTMs encounters two challenges. First, the separation efficiency should be high enough to enable the separation of very close structures. The second challenge is to avoid protein adsorption onto the silica surface of the capillary, which lowers the separation efficiency and leads to analyte loss. The most common approach to reduce protein adsorption in CE-MS is the modification of the capillaries by using permanent or semi-permanent coating materials. This has been already discussed in Section 5.2.2.2. This part focuses on new developments for the analysis of PTMs of intact proteins by CE-MS since 2010 and is divided according to the concerned PTM. For all cited examples in this part, the combination of coated capillaries (either neutral or cationic) with acidic BGE (acetic or formic acids) allowed the separation of the different proteoforms. Indeed, those conditions offer a lot of advantages, such as limitation of protein adsorption and control of EOF, enabling the migration of isoforms according to their own electrophoretic mobilities.

### 5.5.1.1 Glycosylation

Glycosylation is the common PTM, and it corresponds to the attachment of an oligosaccharidic chain to either an asparagine for N-glycosylation or a serine or threonine for O-glycosylation. Glycosylation plays a critical role in a wide range of biological processes, such as inter- and intracellular signaling, inflammation, protein folding, and protein stabilization [104]. Regarding glycosylation, it is mainly investigated either to characterize biopharmaceuticals or to identify abnormal glycoforms as potential disease biomarkers.

Biopharmaceuticals or biologic drugs have gained increasing importance in the drug market. Over 40% of approved biopharmaceuticals are glycoproteins, so the characterization of this PTM is of great importance to assess the quality of biopharmaceuticals. Indeed, glycosylation is a major source of protein heterogeneity and can affect clinical safety and efficacy profiles [105].

In 2011, Taichrib et al. [106] developed a CZE method for the detection of isotope distributions of intact erythropoietin (EPO), using a neutrally coated capillary. The high decrease of the EOF allowed to partially separate not only protein glycoforms differing by the number of sialic acids but also the number of neutral hexoses. Thereby, isoforms with a mass difference of 1% could be separated based on differences in electrophoretic mobility. The isoforms were detected with a high-resolution TOF mass spectrometer (Rs > 40,000), enabling the detection of isotope distributions of partially overlapping glycoforms.

Moreover, as some biopharmaceuticals come off patent, a high number of biosimilars have appeared. In order to compare different biosimilars of EPO with the original molecule, Taichrib et al. developed, in 2012 [63], a multivariate statistics tool based on the CZE-ESI-MS separation method that they previously developed in 2011, using a neutrally coated capillary (Ultratrol™ LN). Thanks to this neutral coating and an acidic buffer (1M acetic acid), EOF was suppressed to near zero, so glycoforms were separated by their own mobility only (i.e., based on the number of sialic acids and the number of HexHexNAc units). They applied different chemometrical approaches for the data analysis of glycoform profiles. Relative peak areas of selected intact EPO isoforms were used as variables in principal component analysis and hierarchical agglomerative clustering. Both approaches were suited for the clear differentiation of all EPO preparations differing by the manufacturer, production cell line, or batch number.

Haselberg et al. investigated a triple layer of polybrene and dextran PB-DS-PB for the CZE-ESI-MS separation of recombinant human interferon-β-1a (rhIFN-β) glycoform heterogeneity [107]. This method allowed the separation of 10 glycoforms, depending on their number of sialic acids or neutral hexoses, where 5 out of the 6 glycoforms described in the *European Pharmacopoiea* monograph of IFN-β [108] were observed. Each glycoform could be quantified, enabling a quality control of rhIFN-β batches. In 2013, the same group developed a new method by using a porous tip sprayer as a new sheathless CE-ESI-MS interface and a polyacrylamide coated capillary to increase sensitivity [107], allowing the increase of detection up to 18 glycoforms, using 10 times lower concentration of rhIFN-β (Figure 5.11). The gain in glycoform coverage obtained with this sheathless system was because of

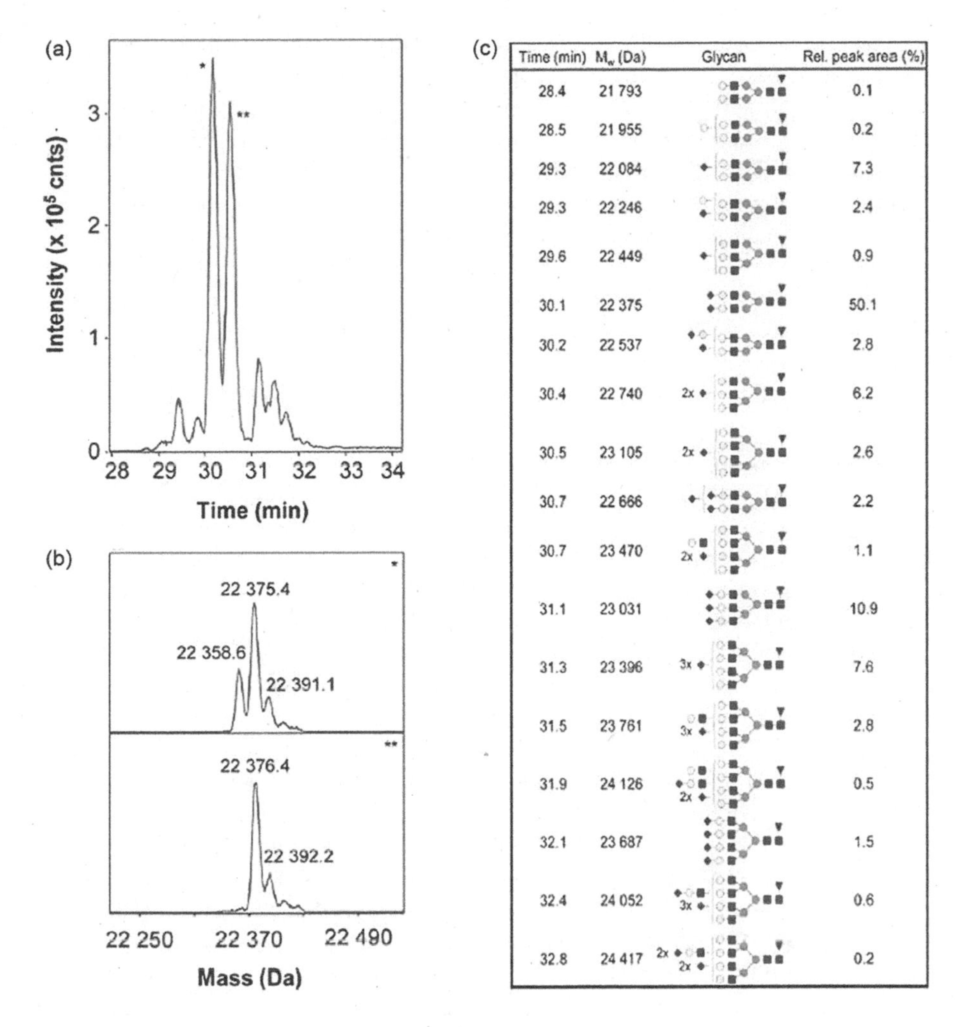


| Time (min) | $M_w$ (Da) | Glycan | Rel. peak area (%) |
|---|---|---|---|
| 28.4 | 21 793 | | 0.1 |
| 28.5 | 21 955 | | 0.2 |
| 29.3 | 22 084 | | 7.3 |
| 29.3 | 22 246 | | 2.4 |
| 29.6 | 22 449 | | 0.9 |
| 30.1 | 22 375 | | 50.1 |
| 30.2 | 22 537 | | 2.8 |
| 30.4 | 22 740 | 2x | 6.2 |
| 30.5 | 23 105 | 2x | 2.6 |
| 30.7 | 22 666 | | 2.2 |
| 30.7 | 23 470 | 2x | 1.1 |
| 31.1 | 23 031 | | 10.9 |
| 31.3 | 23 396 | 3x | 7.6 |
| 31.5 | 23 761 | 3x | 2.8 |
| 31.9 | 24 126 | 2x | 0.5 |
| 32.1 | 23 687 | | 1.5 |
| 32.4 | 24 052 | 3x | 0.6 |
| 32.8 | 24 417 | 2x 2x | 0.2 |

**FIGURE 5.11** (a) BPE obtained during sheathless CE-ESI-MS of rhIFN-β (45 μg/mL) employing a neutrally coated capillary. (b) Deconvoluted mass spectra obtained in the apexes of the peaks migrating at 30.1 min (*) and 30.5 min (**). (c) Migration time, molecular mass, N-glycan composition and structure, and relative peak area for rhIFN-β glycoforms, as observed with sheathless CE-ESI-MS employing a neutrally coated capillary. Symbols: green circle/yellow circle, hexose (mannose/galactose); red triangle, fucose; blue square, N-acetylhexosamine; and purple diamond, sialic acid. Conditions: BGE, 50 mM acetic acid (pH 3.0); CE voltage, 15 kV; tip-to-end plate distance, 1.0 mm. (Reprinted from Haselberg, R. et al., *Anal. Chem.*, 85, 2289–2296, 2013. Copyright 2013 American Chemical Society. With permission.)

both the improved resolution obtained with neutral coated capillaries and the high nano-ESI sensitivities provided by this porous tip interface. This new system also enables to improve the LOD (in the picomolar range). A more efficient glycoprofiling of recombinant human EPO (rhEPO) was also observed (separation of 74 glycoforms). Bush et al. [109] reported a CZE coupled via a sheathless CZE-ESI-MS interface to an Orbitrap ELITE MS for the intact analysis of recombinant human interferon-β1 (Avonex, rhIFN-β1). Using a positively charged PEI coating, column efficiencies between 350,000 and 450,000 plates were produced, allowing separation based on charge and subtle hydrodynamic volume differences. An acidic buffer (3% acetic acid, pH 2.5) was used to protonate the various proteoforms for favorable CZE separation and ESI ionization. A total of 138 proteoforms were found and 55 were quantitated. The separation of isobaric positional isomers such as isomers with the sialic acid residue on either the α(1–3) or the α(1–6) galactose antenna was achieved.

The combination of CZE and ESI-MS was also employed for the analysis of intact glycoproteins in biological samples for the study of the potential of protein glycoforms as cancer biomarkers. Indeed, Ongay et al. [73] investigated the use of a positively charged coated capillary (DMA-EPyM) for the analysis of alpha acid-1-glycoprotein (AGP). The concentration of this plasma glycoprotein is affected during inflammation. This study focused on the characterization of AGP in serum samples and, for the first time, enabled clinical studies on detailed isoform distribution of intact glycoproteins. By using a coated capillary showing anodal EOF at low pH values (BGE of 1M acetic acid), this method allowed the separation of glycoforms, in reverse polarity, based on AGP variants charge or sialic acid differences and conduced to the separation of more than 150 proteoforms. This method was applied to 16 serum samples and revealed the high variability of human AGP expression as well as some changes in the relative proportions of the observed glycoforms. In the same year, this group tested [72] different statistical techniques such as analysis of variance (ANOVA), principal components analysis (PCA), linear discriminant analysis (LDA), and partial least-squares discriminant analysis (PLS-DA) to investigate AGP isoforms as potential biomarkers in bladder cancer, using the previously developed CZE-ESI-MS method. They compared the data obtained by both CZE-UV and CZE-ESI-TOF-MS. It appeared that CZE-UV did not provide statistically different variables allowing group distinction (healthy/cancer). However, CZE-ESI-TOF-MS was found optimal to point out differences between the two groups based on the composition of glycoforms of intact AGP. Statistical differences between both groups (healthy and cancer) were observed mainly because of the presence of tri- and tetra-antennary fucosylated glycoforms in higher abundance in cancer patients. The combination of CZE with this statistical tool showed the usefulness of CZE-MS for the study of potential biomarkers.

Very recently, CZE-ESI-MS on an intact mAb performed in an acidified methanol-water BGE on a capillary with a positively charged coating (M7C4I) coupled to an Orbitrap mass spectrometer and using a sheathless interface was reported. Under denaturing conditions, baseline separation of the 2X-glycosylated, 1X-glycosylated, and aglycosylated populations as a result of hydrodynamic volume differences was obtained. The presence of a trace quantity of dissociated light chain was also detected in the intact protein analysis [110].

#### 5.5.1.2 Acetylation

Acetylation is an in vivo enzymatic PTM playing an important role in the synthesis, stability, and localization of proteins and corresponds to the addition of a $COCH_3$ functional group to the N-terminus aminoacid or to lysines. Concerning the characterization by CE-MS of intact proteins containing acetylated forms, which differ only from 42 Da and one positive charge, two different strategies were proposed. First, positively charged coatings for the separation of acetylated forms have been widely investigated. From all articles, it seems that those cationic coatings were suitable for the study of samples with a simplified matrix (after an reverse phase liquid chromatography (RPLC) prefractionation, for instance) for very fast separations.

Indeed, Haselberg et al. [69] tested the performance of PB-DS-PB coating by working on a test mixture of in vitro acetylated lysozyme, a model protein where six lysines and the N-terminus can be acetylated. The separation revealed that the mixture contained unmodified lysozyme and seven other acetylated lysozyme forms at different degrees (1–7 acetylations). It showed that this positive coating combined with an acidic buffer (50 mM acetic acid, pH 3) was suitable for the separation of highly closed protein species.

In 2014, Han et al. [111] used a cationic coating composed of PEI for the top-down characterization of *Pyrococcus furiosus* proteome by sheathless CZE-ESI-MS after an RPLC prefractionation to reduce the sample complexity. To increase sample loading up to 14% of the capillary, transient ITP was performed using 50 mM ammonium acetate, pH 4.5, while the separation was achieved with a BGE composed of 0.1% acetic acid and 20% isopropanol. This work was the first one to reveal PTMs of *P. furiosus* proteome such as disulfide bonds and acetylation at the N-terminus (representing 4% of the proteins identified).

In 2013, Sarg et al. [112] used an M7C4I-coated capillary (positively charged) to develop a fast sheathless CZE-ESI-MS method as an alternative method to hydrophilic interaction liquid chromatography (HILIC) to study acetylation and methylation of intact histones H4 (highly alkaline protein found in cell nuclei). Contrary to HILIC, the separation was performed depending solely on the acetylation state of histone into its non-, mono-, di-, tri-, and tetra-acetylated forms in less than 18 min, using a BGE of 0.1% formic acid. Concerning the methylation, which could occur on lysine or arginine residues, the different forms (non-, mono-, di-, and trimethylated) co-migrated with the different acetylated species. Nevertheless, the selectivity of sheathless CZE-ESI-MS enabled a confident assignment between isobaric modifications (acetylation and trimethylation), without the use of high-resolution MS, because of differences in their migration time. Thereby, it allowed a fast evaluation and quantification of acetylation and methylation status of histone, but modification sites were not highlighted.

The second strategy developed was the use of a neutral coating that reduces and even suppresses EOF, increasing therefore the separation window that allows a higher resolution of more complex samples. The CE-MS system using porous tip ESI emitter and a linear polyacrylamide (LPA)-coated capillary developed by Haselberg et al. [107] enabled the detection of many acetylated forms of rhEPO. In 2014, Zhao et al. [113] showed the use of an LPA neutral coating for the top-down analysis of the *Mycobacterium marinum* secretome after only an acetone precipitation. In 2016, the same group [114] used this LPA coating to perform the largest top-down proteome data set from yeast by CZE-ESI-MS. For both studies, this neutral coating combined with a sheath-liquid, with a low concentration in organic solvent (only 10% methanol with 0.5% formic acid), allowed them to identify PTMs such as acetylation in less than 35 min.

#### 5.5.1.3 Phosphorylation

Phosphorylation is an important mechanism for regulating the activity of enzymes, but most studies in CE-MS are performed on phosphopeptides and not on intact phosphoproteins, which are mostly analyzed by CIEF owing to their isoelectric point (pI) shift. In 2014, Han et al. [115] used a positively charged PEI-coated capillary for the study of the Dam1 complex, a stable heterodecamer composed of 10 subunits that can be phosphorylated. CZE allowed to detect 9 of the 10 protein subunits, using a sheath-liquid and a BGE with the same composition (5% formic acid with 5% isopropanol), whereas LC separation only detected 8 of them using 100-folds more sample. Phosphorylation increases both the total protein mass and the overall negative charge, which change the electrophoretic mobility and allow to separate phosphoforms with different phosphorylation degree. Applying this top-down method, they were able to separate phosphoforms with different phosphorylation stoichiometries (up to 5 phophoryl groups) and to determine either the phosphorylation site or the region of the modification (in case of low-abundance species). Top-down analysis was completed with bottom-up and middle-up MS to have additional information about the position of phosphorylated amino acid in the protein sequence.

#### 5.5.1.4 Sulfonation

O-sulfonation refers to the transfer of the sulfonate group ($SO_3^-$) to tyrosine, serine, or threonine residues. Pattky et al. [116] characterized an unexpected covalent O-sulfonation of threonine residues of a commercial peptide RPRTRLHTHRNR called peptide D3 (selected by mirror phage display against monomeric or low-molecular-weight amyloid-β-peptide [Aβ 1–42]) in CZE-ESI-MS. The high hydrophilicity and basicity of this peptide (pI 12.6) hampered the use of LC-separation methods but was compatible with CE-MS. They compared two types of coatings, either a neutral (acrylamide-based) or a positively charged one (PB), using the same running buffer of a 3:1 (v/v) mixture of 1 M acetic acid and 1 M formic acid. The acrylamide-LN coating (neutral) allowed a fast separation of D3 isoforms (free, with O-sulfonation or with ion pairing), and MS detection allowed the differentiation between ion pairing and single O-sulfonation species. Unlike this, the positively charged coating (polybrene) induced a high counter EOF, which increased analysis time and improved resolution, enabling the determination of isobaric species with single or double O-sulfonation.

#### 5.5.1.5 Citrullination

Very recently, Faserl et al. [117] extended their study on histones to a challenging PTM, that is, citrullination by using a PAA-coated capillary in CZE-ESI-MS. Citrullination corresponds to the replacement of arginine by a citrulline (presence of an oxygen atom instead of the imine group in arginine). This conversion adds 0.984 Da to the mass of the intact protein. The separation was performed in acidic conditions, and an external pressure of 1.0 or 1.5 psi was applied at the inlet end. The analysis of intact histone H4 showed a heterogeneity that was not explained by acetylation or methylation, as previously [112], but was because of the replacement of the charge of arginine by a new uncharged citrulline amino acid. Thereby, the non-, mono-, di-, and trimethylated forms co-migrated with the different acetylated and citrullinated species. The selectivity of this separation enables the differentiation of species differing by only a mass of 1 Da.

### 5.5.2 Chemical Degradations of Proteins

Sources of heterogeneity of therapeutic proteins include not only a variety of glycoforms and phosphoforms, as previously discussed in this chapter, but also disulfide bond scrambling or sequence mutations and chemical modifications such as oxidation, deamidation, isomerization, and N- and C-terminal alteration or truncation. Most often, these latter forms are structurally very close to the native one and may be difficult to detect. For instance, deamidation on an asparagine (Asn) residue increases the protein mass by only 0.98 Da, but most important, it introduces one additional negative charge to the protein, owing to the conversion of the Asn residue into aspartic (Asp) or isoaspartic (isoAsp) acid. In contrast, isomerization of Asp into isoAsp acid changes neither the mass nor the local charge of the residue. Detailed characterization and quantitation of these proteoforms are important, as specific modifications can affect the efficacy and safety of the drug [118]. For example, oxidation has been linked to protein aggregation, which has potential for immunogenicity [119–121]. Depending on the site of action, deamidation can have functional implications. It has been reported that the formation of a stable succinimide intermediate at Asn55 in the CDR2 region of IgG1 high chain led to a 70% drop in the drug potency [122].

In 2007, Catai et al. [123] demonstrated the possibility of using a noncovalent polymeric bilayer PB-PVS coating for CE-MS analyses of proteins under alkaline conditions. The developed CZE-ESI-MS method used an ion-trap mass spectrometer equipped with an ESI ion source *via* a coaxial sheath-liquid interface and was applied to the characterization of thermally stressed or aged samples of recombinant human growth hormone (rhGH) [68]. The CZE-ESI-MS analysis of thermally degraded rhGH (somatropin CRS), using a BGE of 75 mM ammonium formate (pH 8.5), revealed the formation of two additional species (peaks 2 and 3 of Figure 5.12a). The mass spectra obtained for the detected peaks (Figure 5.12b–d) showed that the degradation products had virtually the same mass as rhGH (22,124 Da). These results suggest that the additional peaks are desamido forms of rhGH. Indeed, mono- and di-deamidation add only 1 or 2 Da, respectively, to the total protein mass. Detection of such a mass difference would have required a mass spectrometer with a higher resolution than the ion trap.

The performances of CZE-ESI-MS systems employing PB-PVS- and PB-DS-PB-coated capillaries for the characterization of therapeutic proteins (rhGH and IFN-β) subjected to prolonged storage, heat exposure, and/or different pH values were investigated by Haselberg et al. [67]. The acidic protein rhGH could be analyzed using a negatively charged PB-PVS-coated capillary and a medium-pH BGE, whereas a positively charged PB-DS-PB-coated capillary in combination with a low-pH BGE appeared to be very suitable for the analysis of rhIFN-β, a basic protein. Modifications could be assigned based on accurate masses, as obtained with TOF-MS, and migration times with respect to the parent compound. For heat-exposed rhGH, oxidations, sulfonate formation, and deamidations were observed.

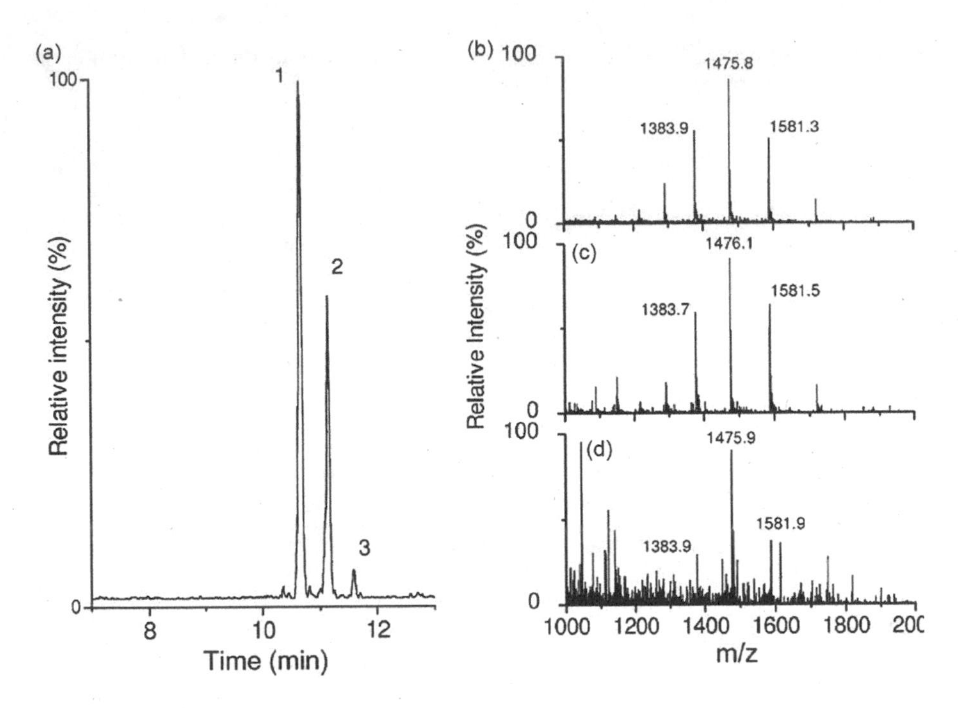


**FIGURE 5.12** CZE-ESI-MS of an rhGH exposed to heat (40°C) for 96 h. A PB-PVS coating was used with a 75 mM ammonium formate pH 8.5 buffer. (a) Sum of extracted ion electropherograms obtained at *m/z* 1475.9 and 1581.5. (b–d) Average mass spectra of peak 1 (intact rhGH), 2, and 3, respectively, which correspond to degradation products. (Reprinted from Catai, J.R. et al., Analysis of recombinant human growth hormone by capillary electrophoresis with bilayer-coated capillaries using UV and MS detection, *J. Chromatog. B*, 852, 160–166, Copyright 2007, with permission from Elsevier.)

In 2013, the same group explored a sheathless CE-MS interface, employing capillaries with a porous tip ESI emitter and an LPA neutral coating for the analysis of intact proteins [107]. Crucial interfacing parameters, such as ESI voltage and ESI tip-to-end plate distance, were optimized for very low flow rates (~5 nL/min) in order to attain maximum sensitivity and stable performance. Analysis of a sample of rhIFN-β allowed the assignment of a variety of deamidation, succinimide, and oxidation products, representing a considerable improvement over sheath-liquid CE-MS.

Besides rhIFN-β1 glycan proteoform separation (described in Section 5.5.1), the sheathless CZE-ESI-MS developed by Bush et al. [109] allowed the detection of its oxidation and deamidation. In addition to deamidation, an associated succinimide was found based on accurate mass. A truncated form with an N-terminal methionine loss (−131 Da mass difference) was also observed. The results could highlight very interesting features, showing, for instance, potential correlations between the methionine loss and the amount of deamidation, as well as between the level of deamidation and the glycan structure. However, the authors stressed that when multiple PTMs and numerous glycan structures exist on a complex glycoprotein, structures corresponding to similar molecular masses can exist; this can potentially lead to mis-assignment. High-resolution separation coupled to high-resolution MS and the ability to conduct top-down $MS^2$ are in these cases of high importance for the characterization of protein heterogeneities.

Recently, Neusüß's group developed 2D capillary electrophoretic techniques coupled to MS for a highly precise mass detection of charge variants. This is an elegant approach to hyphenate ESI-interfering buffer systems (allowing high resolution) online to an MS-compatible second separation dimension. A mechanical four-port nanoliter valve previously introduced [124] was used as the interface between the first and second dimensions. A heart-cut CZE-CZE-MS method has been developed; it allowed the detection of intact model antibody deamidation [34]. Related mass differences (deviation 0.4–0.8 Da) were obtained for glycosylated and deglycosylated variants of trastuzumab. mAb charge variants were first separated in a fused silica capillary by using an electrolyte composed of 380 mM EACA, 1.9 mM triethylenetetramine, and 0.05% w/w (hydroxypropyl)methyl cellulose at pH 5.7. These analytes were then separated in a PVA-coated capillary, using 2M acetic acid as BGE, before reaching directly the mass spectrometer. Another paper reported an iCIEF-CZE-MS developed for the analysis of therapeutics. This method resulted in less contamination to MS and thus improved the spectra quality, thus enabling the detection of accurate masses for not only the main trastuzumab isoform but also its main acidic and basic variants that are attributed to deamidation and succinimide formation or partial cyclisation of N-terminal glutamic acid [51].

For intrinsic (in vivo natural source) and extrinsic (bioprocess and storage) reasons, therapeutic human serum albumin (HSA) preparations, obtained from plasma fractionation, are very heterogeneous. Native HSA can undergo physical (aggregation and dimerization) or chemical (truncation, oxidation, nitrosylation, and glycation) modifications. The free sulfhydryl residue (SH) of the native albumin is implicated in a great number of oxidation reactions. Furthermore, owing to its long half-time, serum albumin is a plasma protein highly sensitive to glycation. The glycation process is a slow nonenzymatic reaction that initially involves the attachment of a hexose (glucose, galactose, or fructose) or derivatives such as glyoxal and methylglyoxal with free amine groups of essentially lysine and arginine residues. This reaction forms reversible adducts that can subsequently be oxidized, polymerized, or cleaved to give irreversible conjugates, called advanced glycation end products (AGEs). Depending on donor population and also on the fractionation process, commercial albumin preparations may vary in their composition regarding the different degraded or modified forms of HSA, and this heterogeneity is suspected to impact their therapeutic effect. A permanent HPC capillary coating was used to minimize albumin adsorption onto the capillary wall and to avoid ionization suppression and/or ion source contamination by removing coating agents. The BGE consisted of 100 mM ammonium bicarbonate, pH 7.8. The CE-MS experiments were carried out with a sheath-liquid composed of acetonitrile-water (60:40, v/v) containing 1% formic acid. To balance the low EOF resulting from the permanent HPC coating, a continuous hydrodynamic pressure was applied during the CZE-ESI-MS analyses. As generally observed with CZE-ESI-MS experiments using a sheath-liquid interface, the coupling led to a loss of resolution. Still, AGE derivatives of HSA such as HSA + MOLD, HSA + Vesperlysines,

HSA + DOLD, HSA + DGH, and HSA + THP could be detected by this technique. Some multiple AGEs, such as HSA + DOLD + 2-Pentosidine or HSA + Argpyrimid ine + 3-Pentosidine were reported for the first time [74].

The CZE-ESI-MS method developed by Tengattini et al. (described in Section 5.4) was also dedicated to the characterization and integrity assessment of the *Mycobacterium tuberculosis* (MTB) antigens TB10.4 (11 kDa) and Ag85B (31 kDa), which were conjugated with 2-iminomethoxyethyl-mannose (Man-IME) and 2-iminomethoxyethyl-mannose (1–6) mannose (Man(1–6)Man-IME). These neo-glycoconjugates are candidate vaccines against tuberculosis. The MTB antigens were exposed to experimental conditions used for chemical glycosylation, in order to investigate their stability during glycovaccine production. CZE-ESI-MS analyses revealed the presence of several closely related degradation products, including truncated, oxidized, and conformational variants, which were assigned by accurate mass. For both antigens, it was observed that glycoforms with a higher number of conjugated saccharides showed a greater susceptibility for deamidation [70]

## 5.6 CONCLUSION

Since the introduction of CE-MS hyphenation, the specific interfaces required an important number of technical developments. In particular, the improvements of CE-MS interfaces have aimed at reducing the impact of the sheath-liquid by providing the hyphenation with nano-ESI sources, in order to improve the sensitivity of the analysis. Thus, several types of high-sensitivity CE-ESI-MS interfaces have been commercially available, which ensure the accessibility of this technique to a large community. In various research fields, the study of intact proteins appears to be relevant, specially to obtain information regarding not only the primary structure but also the three-dimensional (3D) structure. Also, the analysis of intact proteins does not rely on heavy sample preparation, which can introduce experimental biases and generate, to some extent, chemical degradations of proteins, which are not desirable for quality control of biopharmaceuticals. Therefore, it provides the characterization of proteins in a state as close as possible to the actual sample. As a matter of fact, the analysis of intact proteins is meeting a growing interest, even though it has revealed to be more demanding in terms of analytical performances.

This chapter shows that CE-MS represents a powerful analytical tool for the characterization of intact proteins. Indeed, the conjunction of the unique separation selectivity provided by CE to the outstanding performances of MS allows to study massive intact proteins in a complementary manner compared with liquid chromatography-based analysis. On top of that, one major advantage of CE-MS analysis lies in its ability to separate, according to their slight differences in net charge to hydrodynamic radius ratio, a high number of proteoforms with very close structure originated from a wide variety of PTMs, mutations, or chemical degradations of proteins. These modified forms generally involve a mass difference as low as 1 Da (citrullination and deamidation) or only one charge difference (acetylation). CE-MS analyses not only allowed differentiation between species with different phosphorylation, oxidation. or sulfonation degrees but also enabled the detection of glycoforms differing by one monosaccharide in the glycan chain or as close as two isobaric positional isomers differing by the glycosidic bond (α1-3 or α1-6) between sialic acid and galactose residues. The analysis of these modifications is challenging, especially on the intact protein level, regardless of the total size of the protein. The most challenging application described in this chapter (for the first time) is related to the determination of intact mAb aggregation. A native sheathless CE-ESI-MS enabled the detection of monomer and dimer of Trastuzumab, without artifactual association/dissociation occurrence during the ionization process. The most advanced research work cited here has even demonstrated the possibility of using sheath-liquid CE-ESI-MS to achieve the separation and independent characterization of a therapeutic protein exhibiting only a faint conformational change but not the charge nor the mass. The various applications of CE-MS for the analysis of intact proteins have demonstrated the versatility of this instrumental approach to enrich the level of characterization.

The analysis of intact proteins using CE-MS methods still remains at its early stage. Further challenges need to be addressed, in order to improve the intrinsic performances and the application field of CE-MS analysis for intact proteins. The implementation of a preconcentration step prior to the separation, illustrated currently by micro SPE, LVSS, or ITP, has allowed to improve, in a drastic manner, the sensitivity of the analysis by enabling the injection of larger volumes of sample. The strategies described for protein preconcentration offer various alternatives that are still emerging, and further developments are expected concerning that aspect. The coatings compatible with CE-MS experiments currently remain limited, and the development of novel types of capillary coatings may significantly benefit CE-MS analysis of intact proteins. Therefore, innovative capillary coatings open the potentiality to further improve the resolution of intact proteins' separation or tune differently the selectivity. The stability of the coating is of outmost importance to provide a satisfying separation robustness in addition to optimal compatibility with MS analysis. The electrophoretic mode almost systematically used is zone electrophoresis. While this separation mode already demonstrated to be compatible with high-resolution separation of intact proteins, the implementation of additional modes such as isoelectric focusing and gel electrophoresis could further benefit this type of analysis. Different instrumental developments have been already described to improve the compatibility of CIEF with ESI-MS. Also, the application of specific CE methodologies could be further extended, for example, affinity CE. Thus, affinity CE represents an interesting approach that is able to distinguish the formation of protein-ligand complexes and determine their affinity constants by using electrophoretic mobility changes. Another promising application field of CE-MS is protein analysis in native conditions. Native MS often suffers from the absence of a separation method prior to the MS measurement. The implementation of CE-MS in native conditions could alleviate this limitation, in order to improve the performance, especially regarding sample complexity and broaden the applicability of native MS. In conclusion, the numerous developments involving CE-MS methodologies show the acknowledgment of CE-MS from the scientific community as an appropriate technique for the comprehensive characterization of intact proteins.

The major challenges for further implementation of CE-MS are probably the knowledge transfer and an increased implementation and ease of use of CE-MS instruments.

## REFERENCES


1. von Brocke, A., Nicholson, G., and Bayer, E. 2001. Recent advances in capillary electrophoresis/electrospray-mass spectrometry. *Electrophoresis* 22 (7):1251–1266.
2. Gelpi, E. 2002. Interfaces for coupled liquid-phase separation/mass spectrometry techniques. An update on recent developments. *Journal of Mass Spectrometry* 37 (3):241–253.
3. Stutz, H. 2005. Advances in the analysis of proteins and peptides by capillary electrophoresis with matrix-assisted laser desorption/ionization and electrospray-mass spectrometry detection. *Electrophoresis* 26 (7–8):1254–1290.
4. Maxwell, E.J. and Chen, D.D.Y. 2008. Twenty years of interface development for capillary electrophoresis-electrospray ionization-mass spectrometry. *Analytica Chimica Acta* 627 (1):25–33.
5. Hommerson, P., Khan, A.M., de Jong, G.J., and Somsen, G.W. 2011. Ionization thechniques in capillary electrophoresis-mass spectrometry: Principles, design, and application. *Mass Spectrometry Reviews* 30 (6):1096–1120.
6. Bonvin, G., Schappler, J., and Rudaz, S. 2012. Capillary electrophoresis-electrospray ionization-mass spectrometry interfaces: Fundamental concepts and technical developments. *Journal of Chromatography A* 1267:17–31.
7. Zhong, X., Zhang, Z., Jiang, S., and Li, L. 2014. Recent advances in coupling capillary electrophoresis-based separation techniques to ESI and MALDI-MS. *Electrophoresis* 35 (9):1214–1225.
8. Kleparnik, K. 2015. Recent advances in combination of capillary electrophoresis with mass spectrometry: Methodology and theory. *Electrophoresis* 36 (1):159–178.
9. Týčová, A., Ledvina, V., and Klepárník, K. 2017. Recent advances in CE-MS coupling: Instrumentation, methodology, and applications. *Electrophoresis* 38 (1):115–134.
10. De Jong, G.J. 2016. *Capillary Electrophoresis—Mass Spectrometry (CE-MS) Principles and Applications*. Weinheim, Germany: Wiley-VCH.
11. Olivares, J.A., Nguyen, N.T., Yonker, C.R., and Smith, R.D. 1987. Online mass spectrometric detection for capillary zone electrophoresis. *Analytical Chemistry* 59 (8):1230–1232.
12. Bonvin, G., Rudaz, S., and Schappler, J. 2014. In-spray supercharging of intact proteins by capillary electrophoresis–electrospray ionization–mass spectrometry using sheath liquid interface. *Analytica Chimica Acta* 813:97–105.
13. Foret, F., Thompson, T.J., Vouros, P. et al. 1994. Liquid sheath effects on the separation of proteins in capillary electrophoresis electrospray mass spectrometry. *Analytical Chemistry* 66 (24):4450–4458.
14. Haselberg, R., Ratnayake, C.K., de Jong, G.J., and Somsen, G.W. 2010. Performance of a sheathless porous tip sprayer for capillary electrophoresis–electrospray ionization-mass spectrometry of intact proteins. *Journal of Chromatography A* 1217 (48):7605–7611.
15. Haselberg, R., de Jong, G.J., and Somsen, G.W. 2007. Capillary electrophoresis–mass spectrometry for the analysis of intact proteins. *Journal of Chromatography A* 1159 (1–2):81–109.
16. Haselberg, R., de Jong, G.J., and Somsen, G.W. 2011. Capillary electrophoresis-mass spectrometry for the analysis of intact proteins 2007–2010. *Electrophoresis* 32 (1):66–82.
17. Haselberg, R., de Jong, G.J., and Somsen, G.W. 2013. CE-MS for the analysis of intact proteins 2010–2012. *Electrophoresis* 34 (1):99–112.
18. Wojcik, R., Dada, O.O., Sadilek, M., and Dovichi, N.J. 2010. Simplified capillary electrophoresis nanospray sheath-flow interface for high efficiency and sensitive peptide analysis. *Rapid Communications in Mass Spectrometry* 24 (17):2554–2560.

19. Sun, L.L., Knierman, M.D., Zhu, G.J., and Dovichi, N.J. 2013. Fast top-down intact protein characterization with capillary zone electrophoresis-electrospray ionization tandem mass spectrometry. *Analytical Chemistry* 85 (12):5989–5995.
20. Peuchen, E.H., Zhu, G., Sun, L., and Dovichi, N.J. 2017. Evaluation of a commercial electro-kinetically pumped sheath-flow nanospray interface coupled to an automated capillary zone electrophoresis system. *Analytical and Bioanalytical Chemistry* 409 (7):1789–1795.
21. Li, J.J., Wang, C., Kelly, J.F., Harrison, D.J., and Thibault, P. 2000. Rapid and sensitive separation of trace level protein digests using microfabricated devices coupled to a quadrupole-time-of-flight mass spectrometer. *Electrophoresis* 21 (1):198–210.
22. Ramsey, R.S. and Ramsey, J.M. 1997. Generating electrospray from microchip devices using electroosmotic pumping. *Analytical Chemistry* 69 (6):1174–1178.
23. Foret, F. and Kusy, P. 2006. Microfluidics for multiplexed MS analysis. *Electrophoresis* 27 (24):4877–4887.
24. Lazar, I.M., Grym, J., and Foret, F. 2006. Microfabricated devices: A new sample introduction approach to mass spectrometry. *Mass Spectrometry Reviews* 25 (4):573–594.
25. Koster, S. and Verpoorte, E. 2007. A decade of microfluidic analysis coupled with electrospray mass spectrometry: An overview. *Lab on a Chip* 7 (11):1394–1412.
26. Kašička, V. 2016. Recent developments in capillary and microchip electroseparations of peptides (2013–middle 2015). *Electrophoresis* 37 (1):162–188.
27. Mellors, J.S., Gorbounov, V., Ramsey, R.S., and Ramsey, J.M. 2008. Fully integrated glass microfluidic device for performing high-efficiency capillary electrophoresis and electrospray ionization mass spectrometry. *Analytical Chemistry* 80 (18):6881–6887.
28. Mellors, J.S., Black, W.A., Chambers, A.G. et al. 2013. Hybrid capillary/microfluidic system for comprehensive online liquid chromatography-capillary electrophoresis-electrospray ionization-mass spectrometry. *Analytical Chemistry* 85 (8):4100–4106.
29. Redman, E.A., Batz, N.G., Mellors, J.S., and Ramsey, J.M. 2015. Integrated microfluidic capillary electrophoresis-electrospray ionization devices with online MS detection for the separation and characterization of intact monoclonal antibody variants. *Analytical Chemistry* 87 (4):2264–2272.
30. Redman, E.A., Mellors, J.S., Starkey, J.A., and Ramsey, J.M. 2016. Characterization of intact antibody drug conjugate variants using microfluidic capillary electrophoresis-mass spectrometry. *Analytical Chemistry* 88 (4):2220–2226.
31. Zhong, X.F., Maxwell, E.J., and Chen, D.D.Y. 2011. Mass transport in a micro flow-through vial of a junction-at-the-tip capillary electrophoresis-mass spectrometry interface. *Analytical Chemistry* 83 (12):4916–4923.
32. Biacchi, M., Gahoual, R., Said, N. et al. 2015. Glycoform separation and characterization of cetuximab variants by middle-up off-line capillary zone electrophoresis-UV/electrospray ionization-MS. *Analytical Chemistry* 87 (12):6240–6250.
33. Busnel, J.-M., Josserand, J., Lion, N., and Girault, H.H. 2009. Iontophoretic fraction collection for coupling capillary zone electrophoresis with matrix-assisted laser desorption/ionization mass spectrometry. *Analytical Chemistry* 81 (10):3867–3872.
34. Jooß, K., Hühner, J., Kiessig, S., Moritz, B., and Neusüß, C. 2017. Two-dimensional capillary zone electrophoresis–mass spectrometry for the characterization of intact monoclonal antibody charge variants, including deamidation products. *Analytical and Bioanalytical Chemistry* 409 (26):6057–6067.
35. Lee, E.D., Muck, W., Henion, J.D., and Covey, T.R. 1988. Online capillary zone electrophoresis ion spray tandem mass spectrometry for the determination of dynorphins. *Journal of Chromatography* 458:313–321.
36. Lee, E.D., Muck, W., Henion, J.D., and Covey, T.R. 1989. Liquid junction coupling for capillary zone electrophoresis ion spray mass spectrometry. *Biomedical and Environmental Mass Spectrometry* 18 (9):844–850.

37. Fanali, S., D'Orazio, G., Foret, F., Kleparnik, K., and Aturki, Z. 2006. On-line CE-MS using pressurized liquid junction nanoflow electrospray interface and surface-coated capillaries. *Electrophoresis* 27 (23):4666–4673.
38. Wilm, M. and Mann, M. 1996. Analytical properties of the nanoelectrospray ion source. *Analytical Chemistry* 68 (1):1–8.
39. Bonvin, G., Veuthey, J.L., Rudaz, S., and Schappler, J. 2012. Evaluation of a sheathless nanospray interface based on a porous tip sprayer for CE-ESI-MS coupling. *Electrophoresis* 33 (4):552–562.
40. Gahoual, R., Busnel, J.M., Wolff, P., Francois, Y.N., and Leize-Wagner, E. 2014. Novel sheathless CE-MS interface as an original and powerful infusion platform for nanoESI study: from intact proteins to high molecular mass noncovalent complexes. *Analytical and Bioanalytical Chemistry* 406 (4):1029–1038.
41. Fang, L.L., Zhang, R., Williams, E.R., and Zare, R.N. 1994. Online time-of-flight mass spectrometric analysis of peptides separated by capillary electrophoresis. *Analytical Chemistry* 66 (21):3696–3701.
42. Cao, P. and Moini, M. 1997. A novel sheathless interface for capillary electrophoresis/electrospray ionization mass spectrometry using an in-capillary electrode. *Journal of the American Society for Mass Spectrometry* 8 (5):561–564.
43. Moini, M. 2001. Design and performance of a universal sheathless capillary electrophoresis to mass spectrometry interface using a split-flow technique. *Analytical Chemistry* 73 (14):3497–3501.
44. Moini, M. 2007. Simplifying CE-MS operation. 2. Interfacing low-flow separation techniques to mass spectrometry using a porous tip. *Analytical Chemistry* 79 (11):4241–4246.
45. Faserl, K., Kremser, L., Müller, M., Teis, D., and Lindner, H.H. 2015. Quantitative proteomics using ultralow flow capillary electrophoresis–mass spectrometry. *Analytical Chemistry* 87 (9):4633–4640.
46. Gulersonmez, M.C., Lock, S., Hankemeier, T., and Ramautar, R. 2016. Sheathless capillary electrophoresis-mass spectrometry for anionic metabolic profiling. *Electrophoresis* 37 (7–8):1007–1014.
47. Haselberg, R., Harmsen, S., Dolman, M.E.M. et al. 2011. Characterization of drug-lysozyme conjugates by sheathless capillary electrophoresis–time-of-flight mass spectrometry. *Analytica Chimica Acta* 698 (1–2):77–83.
48. Gasilova, N., Gassner, A.L., and Girault, H.H. 2012. Analysis of major milk whey proteins by immunoaffinity capillary electrophoresis coupled with MALDI-MS. *Electrophoresis* 33 (15):2390–2398.
49. Biacchi, M., Bhajun, R., Saïd, N. et al. 2014. Analysis of monoclonal antibody by a novel CE-UV/MALDI-MS interface. *Electrophoresis* 35 (20):2986–2995.
50. Hühner, J. and Neusüß, C. 2016. CIEF-CZE-MS applying a mechanical valve. *Analytical and Bioanalytical Chemistry* 408 (15):4055–4061.
51. Montealegre, C. and Neusüß, C. 2018. Coupling imaged capillary isoelectric focusing with mass spectrometry using a nanoliter valve. *Electrophoresis* 39 (9–10):1151–1154.
52. Pioch, M., Bunz, S.C., and Neususs, C. 2012. Capillary electrophoresis/mass spectrometry relevant to pharmaceutical and biotechnological applications. *Electrophoresis* 33 (11):1517–1530.
53. Hühner, J., Lämmerhofer, M., and Neusüß, C. 2015. Capillary isoelectric focusing-mass spectrometry: Coupling strategies and applications. *Electrophoresis* 36 (21–22):2670–2686.
54. Tang, Q., Harrata, A.K., and Lee, C.S. 1995. Capillary isoelectric focusing-electrospray mass spectrometry for protein analysis. *Analytical Chemistry* 67 (19):3515–3519.
55. Chartogne, A., Tjaden, U.R., and Van der Greef, J. 2000. A free-flow electrophoresis chip device for interfacing capillary isoelectric focusing on-line with electrospray mass spectrometry. *Rapid Communications in Mass Spectrometry* 14 (14):1269–1274.

56. Foret, F., Muller, O., Thorne, J., Gotzinger, W., and Karger, B.L. 1995. Analysis of protein fraction by micropreparative capillary isoelectric-focusing and matrix assisted laser desoption time-of-flight mass spectrometry. *Journal of Chromatography A* 716 (1–2):157–166.
57. Minarik, M., Foret, F., and Karger, B.L. 2000. Fraction collection in micropreparative capillary zone electrophoresis and capillary isoelectric focusing. *Electrophoresis* 21 (1):247–254.
58. Lechner, M., Seifner, A., and Rizzi, A.M. 2008. Capillary isoelectric focusing hyphenated to single- and multistage matrix-assisted laser desorption/ionization-mass spectrometry using automated sheath-flow-assisted sample deposition. *Electrophoresis* 29 (10):1974–1984.
59. Silvertand, L.H.H., Toraño, J.S., de Jong, G.J., and van Bennekom, W.P. 2009. Development and characterization of cIEF-MALDI-TOF MS for protein analysis. *Electrophoresis* 30 (10):1828–1835.
60. Zhang, Z.C., Wang, J.H., Hui, L.M., and Li, L.J. 2011. Membrane-assisted capillary isoelectric focusing coupling with matrix-assisted laser desorption/ionization-fourier transform mass spectrometry for neuropeptide analysis. *Journal of Chromatography A* 1218 (31):5336–5343.
61. Sánchez-Hernández, L., Montealegre, C., Kiessig, S., Moritz, B., and Neusüß, C. 2017. In-capillary approach to eliminate SDS interferences in antibody analysis by capillary electrophoresis coupled to mass spectrometry. *Electrophoresis* 38 (7):1044–1052.
62. Balaguer, E., Demelbauer, U., Pelzing, M. et al. 2006. Glycoform characterization of erythropoietin combining glycan and intact protein analysis by capillary electrophoresis electrospray time-of-flight mass spectrometry. *Electrophoresis* 27 (13):2638–2650.
63. Taichrib, A., Pioch, M., and Neususs, C. 2012. Toward a screening method for the analysis of small intact proteins by CE-ESI-TOF MS. *Electrophoresis* 33 (9–10):1356–1366.
64. Katayama, H., Ishihama, Y., and Asakawa, N. 1998. Stable capillary coating with successive multiple ionic polymer layers. *Analytical Chemistry* 70 (11):2254–2260.
65. Katayama, H., Ishihama, Y., and Asakawa, N. 1998. Stable cationic capillary coating with successive multiple ionic polymer layers for capillary electrophoresis. *Analytical Chemistry* 70 (24):5272–5277.
66. Catai, J.R., Tervahauta, H.A., de Jong, G.J., and Somsen, G.W. 2005. Noncovalently bilayer-coated capillaries for efficient and reproducible analysis of proteins by capillary electrophoresis. *Journal of Chromatography A* 1083 (1):185–192.
67. Haselberg, R., Brinks, V., Hawe, A., Jong, G.J., and Somsen, G.W. 2011. Capillary electrophoresis-mass spectrometry using noncovalently coated capillaries for the analysis of biopharmaceuticals. *Analytical and Bioanalytical Chemistry* 400 (1):295–303.
68. Catai, J.R., Sastre Toraño, J., Jongen, P.M.J.M., de Jong, G.J., and Somsen, G.W. 2007. Analysis of recombinant human growth hormone by capillary electrophoresis with bilayer-coated capillaries using UV and MS detection. *Journal of Chromatography B* 852 (1–2):160–166.
69. Haselberg, R., de Jong, G.J., and Somsen, G.W. 2010. Capillary electrophoresis-mass spectrometry of intact basic proteins using Polybrene-dextran sulfate-Polybrene-coated capillaries: System optimization and performance. *Analytica Chimica Acta* 678 (1):128–134.
70. Tengattini, S., Domínguez-Vega, E., Temporini, C., Terreni, M., and Somsen, G.W. 2016. Monitoring antigenic protein integrity during glycoconjugate vaccine synthesis using capillary electrophoresis-mass spectrometry. *Analytical and Bioanalytical Chemistry* 408 (22):6123–6132.
71. González, N., Elvira, C., San Román, J., and Cifuentes, A. 2003. New physically adsorbed polymer coating for reproducible separations of basic and acidic proteins by capillary electrophoresis. *Journal of Chromatography A* 1012 (1):95–101.

72. Ongay, S. and Neusüβ, C. 2010. Isoform differentiation of intact AGP from human serum by capillary electrophoresis–mass spectrometry. *Analytical and Bioanalytical Chemistry* 398 (2):845–855.
73. Ongay, S., Neusüβ, C., Vaas, S., Díez-Masa, J.C., and de Frutos, M. 2010. Evaluation of the effect of the immunopurification-based procedures on the CZE-UV and CZE-ESI-TOF-MS determination of isoforms of intact α-1-acid glycoprotein from human serum. *Electrophoresis* 31 (11):1796–1804.
74. Marie, A.-L., Przybylski, C., Gonnet, F. et al. 2013. Capillary zone electrophoresis and capillary electrophoresis-mass spectrometry for analyzing qualitative and quantitative variations in therapeutic albumin. *Analytica Chimica Acta* 800:103–110.
75. François, Y.-N., Biacchi, M., Said, N. et al. 2016. Characterization of cetuximab Fc/2 dimers by off-line CZE-MS. *Analytica Chimica Acta* 908:168–176.
76. Biacchi, M., Said, N., Beck, A., Leize-Wagner, E., and François, Y.-N. 2017. Top-down and middle-down approach by fraction collection enrichment using off-line capillary electrophoresis—Mass spectrometry coupling: Application to monoclonal antibody Fc/2 charge variants. *Journal of Chromatography A* 1498:120–127.
77. Marie, A.-L., Dominguez-Vega, E., Saller, F. et al. 2016. Characterization of conformers and dimers of antithrombin by capillary electrophoresis-quadrupole-time-of-flight mass spectrometry. *Analytica Chimica Acta* 947:58–65.
78. Alvarez-Llamas, G., del Rosario Fernandez de la Campa, M., and Sanz-Medel, A. 2003. Sample stacking capillary electrophoresis with ICP-(Q)MS detection for Cd, Cu and Zn speciation in fish liver metallothioneins. *Journal of Analytical Atomic Spectrometry* 18 (5):460–466.
79. Chamoun, J. and Hagege, A. 2005. Sensitivity enhancement in capillary electrophoresis-inductively coupled plasma-mass spectrometry for metal/protein interactions analysis by using large volume stacking with polarity switching. *Journal of Analytical Atomic Spectrometry* 20 (10):1030–1034.
80. Gysler, J., Mazereeuw, M., Helk, B. et al. 1999. Utility of isotachophoresis–capillary zone electrophoresis, mass spectrometry and high-performance size-exclusion chromatography for monitoring of interleukin-6 dimer formation. *Journal of Chromatography A* 841 (1):63–73.
81. Stutz, H., Bordin, G., and Rodriguez, A.R. 2004. Capillary zone electrophoresis of metal-binding proteins in formic acid with UV- and mass spectrometric detection using cationic transient capillary isotachophoresis for preconcentration. *Electroophoresis* 25 (7–8):1071–1089.
82. Tomlinson, A.J., Benson, L.M., Jameson, S., Johnson, D.H., and Naylor, S. 1997. Utility of membrane preconcentration-capillary electrophoresis-mass spectrometry in overcoming limited sample loading for analysis of biologically derived drug metabolites, peptides, and proteins. *Journal of the American Society for Mass Spectrometry* 8 (1):15–24.
83. Figeys, D., Gygi, S.P., Zhang, Y. et al. 1998. Electrophoresis combined with novel mass spectrometry techniques: Powerful tools for the analysis of proteins and proteomes. *Electrophoresis* 19 (10):1811–1818.
84. Peró-Gascón, R., Pont, L., Benavente, F., Barbosa, J., and Sanz-Nebot, V. 2016. Analysis of serum transthyretin by on-line immunoaffinity solid-phase extraction capillary electrophoresis mass spectrometry using magnetic beads. *Electrophoresis* 37 (9):1220–1231.
85. Pont, L., Benavente, F., Barbosa, J., and Sanz-Nebot, V. 2017. On-line immunoaffinity solid-phase extraction capillary electrophoresis mass spectrometry using Fab´antibody fragments for the analysis of serum transthyretin. *Talanta* 170 (Supplement C):224–232.

86. Jensen, P.K., Harrata, A.K., and Lee, C.S. 1998. Monitoring protein refolding induced by disulfide formation using capillary isoelectric focusing-electrospray ionization mass spectrometry. *Analytical Chemistry* 70 (10):2044–2049.
87. Martinović, S., Berger, S.J., Paša-Tolić, L., and Smith, R.D. 2000. Separation and detection of intact noncovalent protein complexes from mixtures by on-line capillary isoelectric focusing-mass spectrometry. *Analytical Chemistry* 72 (21):5356–5360.
88. Borges-Alvarez, M., Benavente, F., Barbosa, J., and Sanz-Nebot, V. 2010. Capillary electrophoresis/mass spectrometry for the separation and characterization of bovine Cu, Zn-superoxide dismutase. *Rapid Communications in Mass Spectrometry* 24 (10):1411–1418.
89. Borges-Alvarez, M., Benavente, F., Barbosa, J., and Sanz-Nebot, V. 2012. Separation and characterization of superoxide dismutase 1 (SOD-1) from human erythrocytes by capillary electrophoresis time-of-flight mass spectrometry. *Electrophoresis* 33 (16):2561–2569.
90. Pont, L., Benavente, F., Barbosa, J., and Sanz-Nebot, V. 2015. Analysis of transthyretin in human serum by capillary zone electrophoresis electrospray ionization time-of-flight mass spectrometry. Application to familial amyloidotic polyneuropathy type I. *Electrophoresis* 36 (11–12):1265–1273.
91. Nguyen, A. and Moini, M. 2008. Analysis of major protein–protein and protein–metal complexes of erythrocytes directly from cell lysate utilizing capillary electrophoresis mass spectrometry. *Analytical Chemistry* 80 (18):7169–7173.
92. Belov, A.M., Viner, R., Santos, M.R. et al. 2017. Analysis of proteins, protein complexes, and organellar proteomes using sheathless capillary zone electrophoresis—Native mass spectrometry. *Journal of The American Society for Mass Spectrometry* 28 (12):2614–2634.
93. Beck, A., Sanglier-Cianferani, S., and Van Dorsselaer, A. 2012. Biosimilar, biobetter, and next generation antibody characterization by mass spectrometry. *Analytical Chemistry* 84 (11):4637–4646.
94. Fermas, S., Gonnet, F., Sutton, A. et al. 2008. Sulfated oligosaccharides (heparin and fucoidan) binding and dimerization of stromal cell-derived factor-1 (SDF-1/CXCL 12) are coupled as evidenced by affinity CE-MS analysis. *Glycobiology* 18 (12):1054–1064.
95. Przybylski, C., Mokaddem, M., Prull-Janssen, M. et al. 2015. On-line capillary isoelectric focusing hyphenated to native electrospray ionization mass spectrometry for the characterization of interferon-gamma and variants. *Analyst* 140 (2):543–550.
96. Righetti, P.G. and Verzola, B. 2001. Folding/unfolding/refolding of proteins: Present methodologies in comparison with capillary zone electrophoresis. *Electrophoresis* 22 (12):2359–2374.
97. Sergeev, N.V., Gloukhova, N.S., Nazimov, I.V. et al. 2001. Monitoring of recombinant human insulin production by narrow-bore reversed-phase high-performance liquid chromatography, high-performance capillary electrophoresis and matrix-assisted laser desorption ionisation time-of-flight mass spectrometry. *Journal of Chromatography A* 907 (1):131–144.
98. Hapuarachchi, S., Fodor, S., Apostol, I., and Huang, G. 2011. Use of capillary electrophoresis-sodium dodecyl sulfate to monitor disulfide scrambled forms of an Fc fusion protein during purification process. *Analytical Biochemistry* 414 (2):187–195.
99. Bertoletti, L., Bisceglia, F., Colombo, R. et al. 2015. Capillary electrophoresis analysis of different variants of the amyloidogenic protein β2-microglobulin as a simple tool for misfolding and stability studies. *Electrophoresis* 36 (19):2465–2472.

100. Bertoletti, L., Schappler, J., Colombo, R. et al. 2016. Evaluation of capillary electrophoresis-mass spectrometry for the analysis of the conformational heterogeneity of intact proteins using beta2-microglobulin as model compound. *Analytica Chimica Acta* 945:102–109.
101. Mironov, G.G., Clouthier, C.M., Akbar, A., Keillor, J.W., and Berezovski, M.V. 2016. Simultaneous analysis of enzyme structure and activity by kinetic capillary electrophoresis–MS. *Nature Chemical Biology* 12:918.
102. Clouthier, C.M., Mironov, G.G., Okhonin, V., Berezovski, M.V., and Keillor, J.W. 2012. Real-time monitoring of protein conformational dynamics in solution using kinetic capillary electrophoresis. *Angewandte Chemie International Edition* 51 (50):12464–12468.
103. Berezovski, M. and Krylov, S.N. 2002. Nonequilibrium capillary electrophoresis of equilibrium mixtures—A single experiment reveals equilibrium and kinetic parameters of protein–DNA interactions. *Journal of the American Chemical Society* 124 (46):13674–13675.
104. Varki, A. and Lowe, J.B. 2009. *Biological Role of Glycans, Essentials of Glycobiology.* Cold Spring Harbor, NY: Cold Spring Harbor Laboratory Press.
105. Schiestl, M., Stangler, T., Torella, C. et al. 2011. Acceptable changes in quality attributes of glycosylated biopharmaceuticals. *Nature Biotechnology* 29:310.
106. Taichrib, A., Pelzing, M., Pellegrino, C., Rossi, M., and Neusüß, C. 2011. High resolution TOF MS coupled to CE for the analysis of isotopically resolved intact proteins. *Journal of Proteomics* 74 (7):958–966.
107. Haselberg, R., de Jong, G.J., and Somsen, G.W. 2013. Low-flow sheathless capillary electrophoresis-mass spectrometry for sensitive glycoform profiling of intact pharmaceutical proteins. *Analytical Chemistry* 85 (4):2289–2296.
108. European Pharmacopoiea. Supplement 6.8 2010. *Interferon beta-1a concentrated solution (sixth ed.).*
109. Bush, D.R., Zang, L., Belov, A.M., Ivanov, A.R., and Karger, B.L. 2016. High resolution CZE-MS quantitative characterization of intact biopharmaceutical proteins: Proteoforms of interferon-β1. *Analytical Chemistry* 88 (2):1138–1146.
110. Belov, M.A., Zang, L., Sebastiano, R. et al. 2018. Complementary middle-down and intact monoclonal antibody proteoform characterization by capillary zone electrophoresis—Mass spectrometry. *Electrophoresis* 30 (16):2069–2082.
111. Han, X., Wang, Y., Aslanian, A. et al. 2014. Sheathless capillary electrophoresis-tandem mass spectrometry for top-down characterization of pyrococcus furiosus proteins on a proteome scale. *Analytical Chemistry* 86 (22):11006–11012.
112. Sarg, B., Faserl, K., Kremser, L. et al. 2013. Comparing and combining capillary electrophoresis electrospray ionization mass spectrometry and nano-liquid chromatography electrospray ionization mass spectrometry for the characterization of post-translationally modified histones. *Molecular & Cellular Proteomics* 12 (9):2640–2656.
113. Zhao, Y.M., Sun, L.L., Champion, M.M., Knierman, M.D., and Dovichi, N.J. 2014. Capillary zone electrophoresis-electrospray ionization-tandem mass spectrometry for top-down characterization of the mycobacterium marinum secretome. *Analytical Chemistry* 86 (10):4873–4878.
114. Zhao, Y., Sun, L., Zhu, G., and Dovichi, N.J. 2016. Coupling capillary zone electrophoresis to a Q exactive HF mass spectrometer for top-down proteomics: 580 proteoform identifications from yeast. *Journal of Proteome Research* 15 (10):3679–3685.
115. Han, X., Wang, Y., Aslanian, A. et al. 2014. In-line separation by capillary electrophoresis prior to analysis by top-down mass spectrometry enables sensitive characterization of protein complexes. *Journal of Proteome Research* 13 (12):6078–6086.

116. Pattky, M., Nicolardi, S., Santiago-Schübel, B. et al. 2015. Structure characterization of unexpected covalent O-sulfonation and ion-pairing on an extremely hydrophilic peptide with CE-MS and FT-ICR-MS. *Analytical and Bioanalytical Chemistry* 407 (22):6637–6655.
117. Faserl, K., Sarg, B., Maurer, V., and Lindner, H.H. 2017. Exploiting charge differences for the analysis of challenging post-translational modifications by capillary electrophoresis-mass spectrometry. *Journal of Chromatography A* 1498:215–223.
118. Hmiel, L.K., Brorson, K.A., and Boyne, M.T. 2015. Post-translational structural modifications of immunoglobulin G and their effect on biological activity. *Analytical and Bioanalytical Chemistry* 407 (1):79–94.
119. Eon-Duval, A., Broly, H., and Gleixner, R. 2012. Quality attributes of recombinant therapeutic proteins: An assessment of impact on safety and efficacy as part of a quality by design development approach. *Biotechnology Progress* 28 (3):608–622.
120. Cacia, J., Keck, R., Presta, L.G., and Frenz, J. 1996. Isomerization of an aspartic acid residue in the complementarity-determining regions of a recombinant antibody to human IgE: Identification and effect on binding affinity. *Biochemistry* 35 (6):1897–1903.
121. Huang, L., Lu, J., Wroblewski, V.J., Beals, J.M., and Riggin, R.M. 2005. In vivo deamidation characterization of monoclonal antibody by LC/MS/MS. *Analytical Chemistry* 77 (5):1432–1439.
122. Yan, B., Steen, S., Hambly, D. et al. 2009. Succinimide formation at Asn 55 in the complementarity determining region of a recombinant monoclonal antibody IgG1 heavy chain. *Journal of Pharmaceutical Sciences* 98 (10):3509–3521.
123. Catai, J.R., Torano, J.S., de Jong, G.J., and Somsen, G.W. 2007. Capillary electrophoresis-mass spectrometry of proteins at medium pH using bilayer-coated capillaries. *Analyst* 132 (1):75–81.
124. Kohl, F.J., Montealegre, C., and Neusüß, C. 2016. On-line two-dimensional capillary electrophoresis with mass spectrometric detection using a fully electric isolated mechanical valve. *Electrophoresis* 37 (7–8):954–958.

# Index

**Note**: Page numbers in italic and bold refer to figures and tables respectively.